Plant Virus Evolution

Marilyn J. Roossinck
Editor

Plant Virus Evolution

Springer

Dr. Marilyn J. Roossinck
The Samuel Roberts Noble Foundation
Plant Biology Division
P.O. Box 2180
Ardmore, OK 73402
USA

Cover Photo: Integrated sequences of *Petunia vein cleaning virus* (in red) are seen in a chromosome spread of *Petunia hybrida* (see Chapter 4).

ISBN: 978-3-540-75762-7 e-ISBN: 978-3-540-75763-4

Library of Congress Control Number: 2007940847

Cover design: WMXDesign GmbH, Heidelberg, Germany

Printed on acid-free paper

9 8 7 6 5 4 3 2 1

springer.com

Preface

The evolution of viruses has been a topic of intense investigation and theoretical development over the past several decades. Numerous workshops, review articles, and books have been devoted to the subject. Medical practitioners have recognized the importance of viral evolution when treating patients with viral diseases. Farmers have recognized the importance of understanding virus evolution in combating emerging viral diseases in their crop plants. As with any field where knowledge is rapidly expanding, many controversies have also arisen about the nature of virus evolution, how to describe virus populations, how to analyze sequence data and estimate phylogenies, etc. Differing points of view will also be found in the various chapters of this book, and I leave it to the readers to decide for themselves which side they find most helpful. In some cases it seems to me that all sides are correct. In other cases, future historians will decide.

This book focuses on the evolution of plant viruses, although some chapters also draw on the more extensive knowledge of animal viruses. It covers topics on evolutionary mechanisms, viral ecology and emergence, appropriate methods for analysis, and the role of evolution in taxonomy. It includes RNA viruses, DNA viruses, integrated viruses and viroids. I hope that this book will provide a much needed reference for all virologists, teachers, plant pathologists, and evolutionists, and that it will inspire young investigators to explore the topic of plant virus evolution in their research. In many cases plant viruses make excellent models for understanding basic principles of evolution, ecology, and animal/human viral evolution. Plant viruses provide experimental systems that cannot be established for animal viruses, such as the generation of unlimited numbers of genetically identical hosts and the inexpensive cultivation and infection of these hosts. Plant viruses were the first viruses discovered, and they have been studied for more than 100 years. With this book plant virology has finally come of age.

October 2007 Marilyn J. Roossinck

Contents

Contributors

Akhtar Ali
Department of Biological Science, University of Tulsa, 600 South College Avenue, Tulsa, OK 74104-3189, USA

Keith A. Crandall
Genoma LLC, Woodland Hills, UT 84653, USA, and Department of Biology, Brigham Young University, Provo, UT 84602, USA

Aurora Fraile
Centro de Biotecnología y Genómica de Plantas and Departamento de Biotecnología, Universidad Politécnica de Madrid, E.T.S.I. Agrónomos, 28040 Madrid, Spain

Fernando García-Arenal
Centro de Biotecnología y Genómica de Plantas and Departamento de Biotecnología, Universidad Politécnica de Madrid, E.T.S.I. Agrónomos, 28040 Madrid, Spain

Robert L. Gilbertson
Department of Plant Pathology, University of California, Davis CA 95616, USA

Anne-Lise Haenni
Institut Jacques Monod, CNRS et Universités Paris 6 et 7, 2 Place Jussieu – Tour 43, 75251 Paris Cedex 5, France, and Grupo Inmunovirologia, Facultad de Medicina, Universidad de Antioquia, Medellin, Colombia

Glyn Harper
John Innes Centre, Norwich Research Park, Colney, Norwich NR4 7UH, UK

Thomas Hohn
Botanical Institute, University of Basel, Schönbeinstrasse 6, 4056 Basel, Switzerland, and Friedrich Miescher Institute for Biomedical Research, Maulbeerstrasse 66, 4058 Basel, Switzerland

Roger Hull
John Innes Centre, Norwich Research Park, Colney, Norwich NR4 7UH, UK

Marie-Line Iskra-Caruana
CIRAD/UMR BGPI 83, 34398 Montpellier Cedex 5, France

Peter D. Nagy
Department of Plant Pathology, 201F Plant Science Building, University of Kentucky, Lexington, KY 40546, USA

Robert A. Owens
Molecular Plant Pathology Laboratory, Beltsville Agricultural Research Center, (USDA/ARS), 10300 Baltimore Blvd., Beltsville, MD 20705, USA

Marcos Pérez-Losada
Genoma LLC, Woodland Hills, UT 84653, USA

Justin S. Pita
Plant Biology Division, The Samuel Roberts Noble Foundation Inc., P.O. Box 2180, Ardmore, OK 73402, USA

Megan Porter
Department of Biological Sciences, University of Maryland, Baltimore County, Baltimore, MD 21250, USA

Alison G. Power
Department of Ecology & Evolutionary Biology, Corson Hall, Cornell University, Ithaca, NY 14853, USA

Katja R. Richert-Pöggeler
Federal Biological Research Centre for Agriculture and Forestry (BBA), Messeweg 11–12, 38104 Braunschweig, Germany

Maria R. Rojas
Department of Plant Pathology, University of California, Davis CA 95616, USA

Marilyn J. Roossinck
Plant Biology Division, The Samuel Roberts Noble Foundation Inc., P.O. Box 2180, Ardmore, OK 73402, USA

Trude Schwarzacher
University of Leicester, University Road, Leicester LE1 7RH, UK

Christina Staginnus
Gregor Mendel Institute of Molecular Plant Biology GmbH, Dr. Bohr-Gasse 3 1030 Vienna, Austria, and GenXPro GmbH, Altenhöferallee 3, 60438 Frankfurt, Germany

Chee How Teo
University of Leicester, University Road, Leicester LE1 7RH, UK

Pierre-Yves Teycheney
CIRAD-UPR75, Station de Neufchâteau, Sainte-Marie, 97130 Capesterre Belle-Eau, Guadeloupe

Chapter 1
Questions and Concepts in Plant Virus Evolution: a Historical Perspective

Fernando García-Arenal(✉) and Aurora Fraile

Abstract The interest in plant virus evolution can be dated to the late 1920s, when it was shown that plant virus populations were genetically heterogeneous, and that their composition changed according to the experimental conditions. Many important ideas were generated prior to the era of molecular virology, such as the role of host- and vector-associated selection in virus evolution, and also that small populations, gene coadaptation and evolutionary trade-offs could limit the efficiency of selection. The analysis of viral genomes in the 1980s and 1990s established the quasispecies-like structure of their populations and allowed extensive analyses of the relationships among virus strains and species. The concept that virus populations had huge sizes and high rates of adaptive mutations became prevalent in this period, with selection mostly invoked as explaining observed patterns of population structure and evolution. In recent times virus evolution has been coming into line with evolutionary biology, and a more complex scenario has emerged. Population bottlenecks during host colonization, during host-to-host transmission or during host population fluctuations may result in smaller population sizes, and genetic drift has been recognized as an important evolutionary factor. Also, particularities of viral genomes such as low levels of neutrality, multifunctionality of coding and encoded sequences or strong epistasis could constrain the plasticity of viral genomes and hinder their response to selection. Exploring the complexities of plant virus evolution will continue to be a challenge for the future, particularly as it affects host, vector and ecosystem dynamics.

Fernando García-Arenal
Centro de Biotecnología y Genómica de Plantas and Departamento de Biotecnología, E.T.S.I. Agronomos, Ciudad Universitaria, Madrid 28040, Spain
fernando.garciaarenal@upm.es

M.J. Roossinck (ed.), *Plant Virus Evolution.*

1.1 Introduction

As is the case with all living entities, reproduction of plant viruses may result in the generation of individuals that differ genetically from their parents, which are called mutants or, more vaguely, variants. Hence, populations of plant viruses are genetically heterogeneous, and the frequency distribution of genetic variants in the population (i.e., the genetic structure of the population) may change with time. This process is called evolution. A major area in the study of evolution aims at understanding the mechanisms of evolution and how they shape the genetic structure of populations. Another area aims at understanding the evolutionary history of organisms and the resulting taxonomic relationships among them. Both aspects of evolutionary studies have a long history in plant virology and have attracted much interest in the last few decades, particularly since the availability of molecular analytical techniques, such as those allowing the rapid determination of nucleotide sequences.

In this chapter we will review how the analysis of plant virus evolution has itself evolved. We do not pretend to make an exhaustive review, but we hope rather to put emphasis on the concepts that have driven the development of the field, illustrated with references to the publications that introduced those concepts or that, in our opinion, best developed them.

1.2 The Early Period

By this, we refer to the period from the origins of plant virology until the widespread use of molecular techniques for nucleic acid analyses. The heterogeneous nature of plant virus populations was evident as early as 1926, by the isolation of symptom variants from areas with atypical symptoms in systemically infected plants (Kunkel 1947) or after biological cloning through single-lesion passage, once necrotic local lesion hosts (i.e., hypersensitive hosts) had been discovered (Holmes 1929). It was also soon perceived that the major components of virus preparations could vary according to the conditions in which the virus was multiplied and passaged. Numerous reports of serial passage experiments including host shifts showed host-associated changes in viral properties, what led to the concept of host adaptation (Yarwood 1979). These observations were interpreted as due to selection in the new conditions. A major concern was whether selection acted on variants present in the original population, or on variants generated under the new conditions. This conceptual dispute was related to a second one about the possibility of obtaining genetically homogeneous virus preparations by single-lesion cloning. Some virologists, particularly Milton Zaitlin, claimed that the frequent appearance of mutants in virus stocks, known from earlier research with *Tobacco mosaic virus* (TMV; Gierer and Mundry 1958), prevented population homogeneity. The reversibility of host adaptation and the first molecular characterization of the phenomenon (Donnis-Keller et al. 1981) supported the hypothesis of host-associated selection of pre-existing variants. Early molecular analyses also showed that continuous generation of mutants prevented genetic homogeneity in single-lesion-derived stocks (García-Arenal et al. 1984). Hence, the confrontation of the two hypotheses was irrelevant, but it promoted research that showed the relevance of selection as an evolutionary process in plant viruses and the intrinsic heterogeneity of plant virus populations.

Evidence that selection could operate rapidly in viral populations also came from natural populations, particularly in relation to the overcoming of resistance factors in crops. The analysis of the selection of pathotype P1 of *Tomato mosaic virus*, which overcomes *Tm-1* gene resistance in tomato, continues to be a classic (Pelham et al. 1970). However, it was also noticed that selection would not always be so effective, as evidenced by the durability of some resistance factors to viruses in crops. Bryan D. Harrison was responsible for three seminal concepts in this respect. He proposed that the evolutionary relevant size of virus populations could approach the number of infected plants or of viruliferous vectors, being thus much smaller than suggested by the high number of virus particles accumulating in the infected plant. Relatively small population sizes could hinder the efficiency of selection in virus populations (Harrison 1981). In addition, his work on *Raspberry ringspot virus* showed two phenomena also limiting the efficiency of selection: selection for mutual compatibility between RNAs 1 and 2 of the virus, and the existence of evolutionary trade-offs, two concepts that became very important in pathogen evolution theory (Hanada and Harrison 1977).

Interest in the evolution of viruses as taxonomic entities (the concept of virus species was slow to be accepted by plant virologists) also originated in this period. Analyses of relatedness among viruses or strains were initially based on biological assays, such as the extent of cross-protection. Later, serological differentiation indices or the amino acid composition of the coat proteins allowed development of quantitative analyses (Van Regenmortel 1975). The work of Adrian Gibbs pioneered the establishment of phylogenetic relationships among plant viruses, and he was also a pioneer in the development of analytical tools, as exemplified by his work on the relationships among the species of tobamoviruses (Gibbs 1986).

Thus, many of the ideas and conceptual approaches relevant to understanding virus evolution, to be developed later on, were generated in this early period on the bases of sound biological experiments or observations, in spite of limited experimental tools.

1.3 The Analysis of Viral Genomes and Its Impact on Virus Evolution Research: Quasispecies and Phylogenetics

The development in the 1970s of methods for the analyses of nucleic acids had a deep impact on the study of virus evolution. These methods allowed the comparison of virus isolates on the basis of genomic regions or viral proteins other than the structural ones, and eventually allowed the comparison of complete genomes.

Comparison of viral variants made much use of ribonuclease T1 fingerprinting, restriction fragment length polymorphisms (RFLPs), ribonuclease protection assay (RPA) of a labeled complementary RNA probe or single-stranded conformation polymorphisms (SSCPs), in addition to nucleotide sequence determination of genomes or parts of genomes. Data from fingerprints, RFLPs and, of course, nucleotide sequences can be used to directly estimate genetic distances between genotypes, while data from RPA and SSCPs cannot, as they depend on sequence context. These methodological limitations often were overlooked because initial analyses of virus variability focused just on the detection of variants, but later handicapped the development of quantitative analyses of population structure.

The availability of methods allowing the differentiation of closely related genotypes, and the availability of biologically active complementary DNA (cDNA) clones of RNA genomes, definitively determined that virus populations are intrinsically heterogeneous owing to errors during replication. Following the trend with animal- and bacteria-infecting viruses, research focused on RNA viruses, and heterogeneity of cDNA-derived populations was initially shown for *Cucumber mosaic virus* (CMV) satellite RNA and for TMV (Aldahoud et al. 1989; Kurath and Palukaitis 1989). It was shown also, initially for *Tobacco mild green mosaic virus* (TMGMV; Rodríguez-Cerezo and García-Arenal 1989), that the frequency distribution of genotypes in virus populations was gamma, with a major genotype plus a set of minor variants newly generated by mutation or kept at a low level by selection. It was shown later on that the shape of this distribution depended on both the virus and the host plant (Schneider and Roossinck 2000, 2001). This genetic structure had been previously reported for RNA viruses infecting bacteria or animals and had been named a quasispecies (Domingo and Holland 1997), as it corresponded to that predicted by Eigen's quasispecies theory, proposed to describe the evolution of an infinite population of asexual replicators at high mutation rate (Eigen and Schuster 1977). The quasispecies concept has been used often in virology as a mere description for genetically heterogeneous virus populations ("swarms" of mutants), with no concern or awareness for further implications, or for the specific conditions required for the quasispecies concept to materialize, as pointed out by Eigen (1996) himself and developed in the next section. Regardless of the limited appreciation of its implications, the quasispecies concept was crucial in making virologists in the 1980s aware of the intrinsic heterogeneity of virus populations, an early discovery that had been overlooked in an era focused on the molecular analyses of viral genomes.

The quasispecies concept assumed high mutation rates for RNA viruses. It was indeed shown with bacteriophages and with lytic viruses infecting mammalian cells that RNA-dependent RNA polymerases lacked a proofreading activity, and had error rates several orders of magnitude higher than DNA-dependent DNA polymerases of large DNA phages or of cellular organisms (on the order of 10^{-4}–10^{-6} per position and replication round; Drake et al. 1998). Because of high mutation rates of RNA viruses and high accumulation levels in host cells, it was concluded that RNA viruses had large and highly diverse populations. As a consequence, viral populations would easily respond to changing selection pressure, and the evolution of high mutation rates would have an adaptive value, allowing the virus to survive in changing environments. This concept became the "dogma" that has presided over analyses of RNA virus evolution for more than two decades since the early 1980s. Challenges to this dogma, coming initially from the plant virus field, will be described in the next section.

Nucleotide sequence determination, and the development of methods for the comparison of distantly related sequences, led to phylogenetic analyses of proteins with a similar function in viruses belonging to different genera. These analyses, first done with RNA-dependent RNA polymerases (Kamer and Argos 1984), allowed the classification of viruses in large groups or "superfamilies" (Koonin and Dolja 1993; Goldbach and de Haan 1994) although the validity of the higher-order comparisons was later seriously questioned (Zanotto et al. 1996). Availability of nucleotide sequences of complete viral genomes showed that phylogenies of different gene families were not congruent and that gene organization within the genomes could vary between viral taxonomic groups that were otherwise related. This could be explained by "reassortment of functional modules of coding and regulatory sequences" (Haseloff et al. 1984) according to the concept of "modular evolution," first proposed for bacteriophages (Botstein 1980). Also, availability of whole genome sequences showed that virus genes were often contained totally or partially within another gene, in a different reading frame. This observation led Adrian Gibbs to propose the very novel concept of de novo generation of genes by "overprinting," and methods to analyze which of the two overlapping genes was the novel one (Keese and Gibbs 1992).

The ease of comparing viral genomes also prompted analyses of the genetic structure of natural populations of plant viruses. Phylogenetic approaches were generally preferred to population genetics ones. Both approaches showed from the early 1990s that virus populations could be structured according to different factors, such as geographic or host origin, and different selection pressures were invoked to explain the observed population structures. Again, Gibbs's work on tymoviruses infecting wild plants (Skotnicki et al. 1993, 1996) was pioneering in this field. Major selection pressures acting on virus genomes were identified in this period. Selection was associated with the need to maintain a functional structure, for instance, in the capsid protein of tobamoviruses (Altschuh et al. 1987) or in noncoding subviral pathogenic nucleic acids such as satellites or viroids (Fraile and García-Arenal 1991; Elena et al. 1991). Host-associated selection, already known from passage experiments, was also invoked to explain population structure, for instance, in *Kennedya yellow mosaic virus* (KYMV; Skotnicki et al. 1996), *Hop stunt viroid* (Kofalvi et al. 1997) or *Barley yellow dwarf virus* (BYDV; Mastari et al. 1998). Evidence of vector-associated selection initially derived from loss of transmissibility upon mechanical passage or vegetative propagation of the virus host (Reddy and Black 1977). Population structure in relation to vector transmission has been analyzed in few instances, mostly with begomoviruses (Harrison and Robinson 1999; Simón et al. 2003) supporting vector-associated selection.

Because most analyses of virus population structure followed a phylogenetic approach and because analytical methods were able to differentiate between closely related variants, the resulting data were interpreted mostly as conforming to the dogma of high genetic diversity of RNA virus populations. However, the genetic diversity of a population does not depend only on the number of genotypes present in the population, but also on the frequency of each genotype and on the genetic distances among them. The few reports that considered these three factors for the analysis of population structure, such as those for solanaceae-infecting tobamoviruses (Rodríguez-Cerezo et al. 1991) showed low population diversities. Later on, analyses of populations of other viruses also provided evidence of low population diversity and, importantly, showed that population diversity did not depend on the nature, RNA or DNA, of the virus genome (García-Arenal et al. 2001). Also, nucleotide diversity in virus genes was not higher for RNA than for DNA plant viruses and, interestingly, diversity values were in the range of those of the genes of their eukaryotic hosts and vectors. Data showed that negative selection was important in keeping low nucleotide diversities and, more important, that the degree of negative selection was not related to the function of the virus protein, at odds with observations on cellular organisms, in which certain classes of proteins are more conserved than other. Multifunctionality of viral proteins could explain these observations (García-Arenal et al. 2001). Hence, evidence showed high genetic conservation, rather than high diversity, for most plant virus populations analyzed.

1.4 The Challenge to the Dogma: Viruses Might Be Not So Variable nor Might Their Populations Be So Big

The fact that plant viral populations did not show the big diversity assumed by the dogma led to the questioning of the two premises on which that dogma rested: high population sizes and high rates of adaptive mutations. As stated, it was proposed as early as 1981 that in spite of high levels of virus accumulation in the infected hosts, population size perhaps would not be so high (Harrison 1981). In fact, the relevant evolutionary parameter is not the census size of the population, but the effective population size, which could be grossly defined as the fraction of the population that passes its genes to the new generation. Expansions and contractions of population size during the virus life cycle, i.e., the occurrence of population bottlenecks, would affect the effective population size. It was first shown for TMV that virus population indeed passed through severe bottlenecks during plant colonization, and that effective sizes of the population that initiates colonization of a new leaf could be as low as units or tens of individuals (Sacristán et al. 2003). Detailed analyses of within-host population structure of *Wheat soil borne mosaic virus* led to similar numbers, derived from a different approach (French and Stenger 2003). It was also shown that severe population bottlenecks occurred both during CMV colonization of new leaves within a plant and during horizontal transmission by aphids to new plants (Li and Roossinck 2004; Ali et al. 2006). Genetic drift can result in the elimination from the population of the fittest genotypes and the accumulation of deleterious mutations, eventually leading to population extinction (i.e., mutational meltdown), as shown experimentally for various RNA viruses, including *Tobacco etch virus* (de la Iglesia and Elena 2007). Mutation accumulation and population extinction was also shown to occur in nature in the TMV population infecting *Nicotiana glauca*, owing to a reduction in the TMV population size caused by coinfection with TMGMV, to our knowledge the only report of mutational meltdown occurring in viral populations in nature (Fraile et al. 1997). Hence, random genetic drift, as opposed to selection, can be an important evolutionary factor for plant viruses, a possibility not contemplated by the quasispecies theory, which is a deterministic model of evolution.

The first years of this century also brought evidence that the high potential to vary of RNA viruses need not result in high adaptability. The spontaneous mutation rate of TMV was determined using a large (804-nt) mutational target in conditions of minimal selection against deleterious mutants (Malpica et al. 2002). Mutation rates were high but slightly less than those previously reported for lytic RNA viruses (0.05–0.1 compared with approximately one mutation per genome and replication round, but see a new estimation of *Vesicular stomatitis virus* (VSV) mutation rate in line with that of TMV in Furió et al. 2005). More importantly, the mutational spectrum for an RNA genome was reported for the first time in this work. A large percentage of mutants were multiple mutants, and about one third of mutations were insertions and deletions, so a large fraction of mutations will be highly deleterious if not immediately lethal. An analysis of the mutational spectrum of VSV also showed that most point mutations were deleterious (Sanjuán et al. 2004a). Thus, the high mutation rate of RNA viruses seems not to have evolved as an adaptive trait facilitating adaptation to new environments (Elena and Sanjuán 2005). Also, epistatic interactions between different site mutations were shown to be strong for VSV (Sanjuán et al. 2004b). Genetic exchange by recombination or reassortment of genomic segments (i.e., sexuality) is another important source of genetic variation in viruses, often with large phenotypic effects such as host switches, host range expansion and is often at the root of the emergency of new viral diseases. A typical example is the role of genetic exchange in the origin of the pandemic of cassava mosaic disease in East Africa (Fargette et al 2006). Genetic exchange also has been shown to be important in the evolution of taxonomic entities (White et al. 1995). Genotypes generated by recombination can be frequent in virus populations, as shown particularly for begomoviruses (Sanz et al. 2000), but also for RNA virus such as *Turnip mosaic virus* (TuMV) (Tan et al. 2004). Last, a recent report has shown the importance of recombination in *Cauliflower mosaic virus* that can make up to 50% of experimental populations. This report also provides the only estimate of recombination rates, 2×10^{-5}–5×10^{-5} per base and replication cycle, i.e., on the order of mutation rates in RNA viruses (Froissart et al. 2005). However, genetic structure of field RNA virus populations often indicates constraints to genetic exchange (Bonnet et al. 2005), and experiments with both DNA and RNA viruses (*Maize streak virus* and CMV, respectively) have shown that heterologous gene combinations are selected against, supporting the coadaptation of gene complexes in viral genomes (Martin et al. 2005; Escriu et al. 2007). Thus, this is also evidence that epistatic interactions would constrain the plasticity of the small genomes of RNA viruses, further limiting their possibility to respond to selection pressures. The high durability of most resistance factors to viruses in crops, despite the common occurrence of resistance-breaking isolates (Harrison 2002; García-Arenal and McDonald 2003), is in agreement with these observations.

Thus, the general view on the evolution of RNA viruses that dominated the scene in the 1980s and 1990s and that we have called the dogma is unsupported by a large body of evidence. Theoretical and experimental challenges of the quasispecies theory have also multiplied (see the exchange in Holmes and Moya 2002 and Domingo 2002). Conditions for application of the quasispecies theory, such as equilibrium conditions, single-peak (or master sequence) fitness landscapes, large values for the product of effective population size and mutation rate, lack of lethal mutants, or asexuality may often not apply to viral populations (Eigen 1996; Wilke 2005), as shown in this section. In addition, the quasispecies theory may be in fact in perfect agreement with standard population genetics (Wilke 2005). Hence, the view is prevailing that virus evolution is not intrinsically different in its processes from that of other living entities, in spite of particularities. One such particularity would be a not well-defined ploidy, derived from the possibility of coinfection of the same cell by different virus particles. A direct consequence of coinfection is that deleterious or lethal mutants may be efficiently complemented by functional genotypes sharing the same cell, which will provide the affected function *in trans*. Complementation may counter the effect of selection (Moreno et al. 1997) and, thus, may have important consequences on virus evolution, for instance, the maintenance in viral populations of more virulent but less fit variants.

1.5 Recent Times: New Concepts and New Challenges

If virus evolution is not intrinsically different in its processes from that of other living entities, viruses may be good experimental systems to test general evolutionary hypotheses (Elena and Lenski 2003). The use of plant viruses to this aim has only started recently, but probably will be a major component of research on plant virus evolution in the near future. Another important recent trend likely to explode in the near future is the consideration of plant virus evolution on a broader context than the virus population itself, incorporating the interaction of viral and host and/or vector populations.

Analyses of the genetic structure of viral populations and viral phylogenies have benefited in recent times from the availability of new computational tools that allow detailed and more informative analyses of sequence data. Examples are tools that implement different methods for the detection of recombination (Posada and Crandall 2001) or of positive selection acting on particular codons of protein-encoding genes (Yang et al. 2000). These tools have recently been applied to different plant virus systems (Moury 2004; Codoñer and Elena 2006). Progress in methods to obtain and analyze phylogenetic or population structure data have resulted also in an increased awareness that their interpretation is often hindered by limited information on the factors that act on virus populations and determine their evolution. Particularly, host- and vector-associated factors act on virus populations in ways probably dependent on properties of the virus itself, such as host and vector range. Thus, there is a need to analyze virus evolution within broader epidemiological and ecological frames. Good examples of integration of ecological and epidemiological data in virus evolution studies are in a series of reports on the evolutionary biology of *Rice yellow mottle virus* (Fargette et al. 2006) and TuMV (Tomitaka and Ohshima 2006). An important motor of research on virus evolution from an ecological perspective is related to the development of transgenic plants with pathogen-derived resistance, and the need to evaluate the risks that their widespread use could have for agricultural and wild ecosystems (Tepfer 2002).

Another important field just starting to be developed is the role in virus evolution of the effects that viruses have on their host and vector populations. Viruses often can be virulent pathogens that harm their hosts. In the context of evolutionary biology, virulence is often defined as the deleterious effects of parasite infection on the host's fitness. As such, virulence affects the population genetics and dynamics of hosts and is, thus, the major factor determining host–pathogen coevolution (Frank 1996). Because virulence is the key property of pathogens, much theoretical work has been done in the last two decades aimed at modeling virulence evolution. Plant viruses are most adequate systems to test the assumptions and predictions of these models, but the study of virulence evolution in plant viruses is largely an unexplored field. Recent interest in this subject is shown by reports that have addressed questions such as the relationship between virulence and virus multiplication, mode of transmission, host adaptation or within-host competition in mixed infections (Escriu et al. 2003; Sacristán et al. 2005; Stewart et al. 2005). Also, viruses may affect the population dynamics of their insect vector. For instance, it has been shown that virus infection has an influence on both the attraction of insect vectors by host plants and on their reproductive potential (Fereres et al. 1999; Jiu et al. 2007). Interestingly, the effect of virus infection on aphid biology may depend on the transmission manner, nonpersistent or persistent (Eigenbrode et al. 2002). Research on virus–vector interactions presently lags behind that on virus–host interactions.

In the past, most analyses of virus evolution focused on agricultural systems, and little work has been done in natural ecosystems. An analysis of the prevalence of five generalist viruses on 21 species of wild plants has shown a selective interaction between viruses and hosts and, more important, that host selectivity is a successful strategy for generalist viruses (Malpica et al. 2006). This result is relevant, as models of virulence evolution predict that pathogens will evolve to specialism, against the evidence that most plant and animal pathogens are generalists (Woolhouse et al. 2001). The observed tendency to specialize could reconcile both views. This report also showed that viruses tended to associate positively in mixed infected hosts (Malpica et al. 2006), which again is relevant, as coinfection of different pathogens may have important consequences for the pathogen, the infected host and for host–pathogen coevolution.

The role of plant viruses in ecosystem dynamics, as it relates to virus evolution, is also an emergent area of research. Virus infection of wild plants often goes unnoticed because it is asymptomatic, and it generally has been considered not to harm the host. However, several reports have shown that virus infection can decrease the fitness of wild plants (Kelly 1994; Friess and Maillet 1996; Maskell et al. 1999). Perhaps the best studied case is the effect of BYDV and *Cereal yellow dwarf virus* infection in wild grasses in California. Prevalence, accumulation and virulence vary according to the host plant species, and have a complex influence on grassland dynamics, mediated by phenomena such as pathogen spillover (an epidemic in one host is affected by transmission from another host population) and pathogen-mediated apparent competition (Power and Mitchell 2004). Virus infection also influences the effects of herbivory, showing the importance of multitrophic interactions in virus ecology (Malmstrom et al. 2006). The relevance of multitrophic interactions had been shown long ago in wild legumes in Australia infected by KYMV, which were less affected by herbivory than noninfected plants (Gibbs 1980). Another outstanding example of complex interactions is shown by an analysis demonstrating that the increased stress tolerance of grasses associated with colonization by endophytic fungi is due to virus infection of the fungus (Márquez et al. 2007). The analysis of virus populations in ecosystems is, thus, uncovering highly complex networks of interactions. How these interactions affect virus evolution should be an important area of research in the upcoming years.

1.6 Final Comments

In the earlier period of plant virology, prior to the advent of molecular techniques for the analysis of viral genomes, research often had a population-oriented perspective, and reflected the very broad biological understanding that characterized the earlier generations of plant virologists. Many important ideas, often to be fully developed in later years, were generated at this time. The impact of molecular virology in the 1980s somewhat detracted from the interest in plant virus population research, and when this interest was renewed later on most plant virologists, even those interested in evolution, had a molecular rather than a populational formation, which conditioned the perception of evolutionary processes. In recent times, plant virus evolution has again attracted population biologists and the study of plant virus evolution is, thus, losing its peculiarities and is coming in line with evolutionary biology at large. As a consequence, new questions are being asked from new perspectives and broader contexts, including the reciprocal complex interactions of virus populations with those of their hosts and vectors. Exploring the complexities of plant virus evolution will certainly keep scientists busy for a long time.

Acknowledgements We want to thank Adrian Gibbs and Santiago F. Elena for their useful comments on a previous version of this chapter. Research on virus evolution in our laboratory is supported by grants AGL2005-01122 to F.G.-A. and AGL2004-03496 to A.F.

References

Aldahoud R, Dawson WO, Jones GE (1989) Rapid, random evolution of the genetic structure of replicating tobacco mosaic virus populations. Intervirology 30:227–233

Ali A, Li H, Schneider WL, Sherman DJ, Gray S, Smith D, Roossinck MJ (2006) Analysis of genetic bottlenecks during horizontal transmission of Cucumber mosaic virus. J Virol 80:8345–8350

Altschuh D, Lesk AM, Bloomer AC, Klug A (1987) Correlation of co-ordinated amino acid substitutions with function in viruses related to tobacco mosaic virus. J Mol Biol 193:693–707

Bonnet J, Fraile A, Sacristán S, Malpica JM, García-Arenal F (2005) Role of recombination in the evolution of natural populations of Cucumber mosaic virus, a tripartite RNA plant virus. Virology 332:359–368

Botstein D (1980) A theory of modular evolution for bacteriophages. Ann N Y Acad Sci 354:484–490

Codoñer FM, Elena SF (2006) Evolutionary relationships among members of the Bromoviridae deduced from whole proteome analysis. Arch Virol 151:299–307

de la Iglesia F, Elena SF (2007) Fitness declines in *Tobacco etch virus* upon serial bottlenecks transfers. J Virol 81:4941–4947

Domingo E (2002) Quasispecies theory in virology. J Virol 76:463–465

Domingo E, Holland JJ (1997) RNA virus mutations and fitness for survival. Annu Rev Microbiol 51:151–178

Donis-Keller H, Browning KS, Clarck JM (1981) Sequence heterogeneity in satellite tobacco necrosis virus RNA. Virology 110:43–54

Drake JW, Charlesworth B, Charlesworth D, Crow JF (1998) Rates of spontaneous mutation. Genetics 148:1667–1686

Eigen M (1996) On the nature of virus quasispecies. Trends Microbiol 4:216–218

Eigen M, Schuster P (1977) The hypercycle. A principle of natural self-organization. Pt.A.: emergence of the hypercycle. Naturwissenchaften 64:541–565

Eigenbrode SD, Ding H, Shiel P, Berger PH (2002) Volatiles from potato plants infected with potato leafroll virus attract and arrest the virus vector, *Myzus persicae* (Homoptera: Aphididae). Proc R Soc Lond Ser B 269:455–460

Elena SF, Lenski RE (2003) Evolution experiments with microorganisms: The dynamic and genetic bases of adaptation. Nat Rev Genet 4:457–469

Elena SF, Sanjuán R (2005) Adaptive value of high mutation rates of RNA viruses: separating causes from consequences. J Virol 79:11555–11558

Elena SF, Dopazo J, Flores R, Diener TO, Moya A (1991) Phylogeny of viroids, viroid-like satellite RNAs, and the viroid-like domain of hepatitis δ virus RNA. Proc Natl Acad Sci USA 88:5631–5634

Escriu F, Fraile A, García-Arenal F (2003) The evolution of virulence in a plant virus. Evolution 57:755–765

Escriu F, Fraile A, García-Arenal F (2007) Constraints to genetic exchange support gene coadaptation in a tripartite RNA virus. PLoS Pathog 3:e8

Fargette D, Konate G, Fauquet C, Muller E, Peterschmitt M, Thresh JM (2006) Molecular ecology and emergence of tropical plant viruses. Annu Rev Phytopathol 44:235–260

Fereres A, Kampmeier GE, Irwin ME (1999) Aphid attraction and preference for soybean and pepper plants infected with Potyviridae. Ann Entomol Soc Am 92:542–548

Fraile A, García-Arenal F (1991) Secondary structure as a constraint on the evolution of a plant viral satellite RNA. J Mol Biol 221:1065–1069

Fraile A, Escriu F, Aranda MA, Malpica JM, Gibbs AJ, García-Arenal F (1997) A century of tobamovirus evolution in an Australian population of *Nicotiana glauca*. J Virol 71:8316–8320

Frank SA (1996) Models of parasite virulence. Q Rev Biol 71:37–78

French R, Stenger DC (2003) Evolution of wheat streak mosaic virus: dynamics of population growth within plants may explain limited variation. Annu Rev Phytopathol 41:199–214

Friess N, Maillet J (1996) Influence of cucumber mosaic virus infection on the intraspecific competitive ability and fitness of purslane (*Portulaca oleraceae*) New Phytol 132:103–111

Froissart R, Roze D, Uzest M, Galibert L, Blanc S, Michalakis Y (2005) Recombination every day: abundant recombination in a virus during a single multi-cellular host infection. PLoS Biol 3:389–395

Furió V, Moya A, Sanjuán R (2005) The cost of replication fidelity in an RNA virus. Proc Natl Acad Sci USA 102:10233–10237

García-Arenal F, McDonald B (2003) An analysis of the durability of resistance of plant to viruses. Phytopathology 93:941–952

García-Arenal F, Palukaitis P, Zaitlin M (1984) Strains and mutants of tobacco mosaic virus are both found in virus derived from single-lesion-passaged inoculum. Virology 132:131–137

García-Arenal F, Fraile A, Malpica JM (2001) Variability and genetic structure of plant virus populations. Annu Rev Phytopathol 39:157–186

Gibbs AJ (1980) A plant virus that partially protects its wild legume host against hebivores. Intervirology 13:42–47

Gibbs AJ (1986) Tobamovirus classification. In: Van Regenmortel MHV, Fraenkel-Conrat H (eds) The plant viruses, vol 2. The rod-shaped plant viruses. Plenum, New York, pp 168–180

Gierer A, Mundry KW (1958) Productions of mutants of tobacco mosaic virus by chemical alteration of its ribonucleic acid *in vitro*. Nature 182:1457–1458

Goldbach R, de Haan P (1994) RNA viral supergroups and the evolution of RNA viruses. In: Morse SS (ed) The evolutionary biology of viruses. Raven, New York, pp 105–119

Hanada K, Harrison BD (1977) Affects of virus genotype and temperature on seed transmission of nepoviruses. Ann Appl Biol 85:79–92

Harrison BD (1981) Plant virus ecology: ingredients, interactions and environment influences. Ann Appl Biol 99:195–209

Harrison BD (2002). Virus variation in relation to resistance-breaking in plants. Euphytica 124:181–192

Harrison BD, Robinson DJ (1999) Natural genomic and antigenic variation in whitefly-transmitted geminiviruses (begomoviruses). Annu Rev Phytopathol 37:369–398

Haseloff J, Goelet P, Zimmern D, Ahlquist P, Dasgupta R, Kaesberg P (1984) Striking similarities in amino acid sequence among nonstructural proteins encoded by RNA viruses that have dissimilar genomic organization. Proc Natl Acad Sci USA 81:4358–4362

Holmes FO (1929) Local lessions in tobacco mosaic. Bot Gaz 87:39–55

Holmes EC, Moya A (2002) Is the quasispecies concept relevant to RNA viruses? J Virol 76:460–462

Jiu M, Zhou XP, Tong L, Xu J, Yang X, Wan FH, Liu SS (2007). Vector-virus mutualism accelerates population increase of an invasive whitefly. PLoS ONE 2:e182

Kamer G, Argos P (1984) Primary structural comparisons of RNA-dependent polymerases from plant, animal and bacterial viruses. Nucleic Acids Res 12:7269–7282

Keese P, Gibbs A (1992) Origins of genes: big bang or continuous creation? Proc Natl Acad Sci USA 89:9889–9893

Kelly SE (1994) Viral pathogens and the advantage of sex in the perennial grass *Anthoxantum odoratum*: a review. Philos Trans R Soc Lond Ser B 346:295–302

Kofalvi SA, Marcos JF, Cañizares MC, Pallás V, Candresse T (1997) Hop stunt viroid (HSVd) sequence variants from *Prunus* species: evidence for recombination between HSVd isolates. J Gen Virol 78:3177–3186

Koonin EV, Dolja VV (1993) Evolution and taxonomy of positive-strand RNA viruses: implications of comparative analysis of amino acid sequences. Crit Rev Biochem Mol Biol 28:375–430

Kunkel LO (1947) Variation in phytopathogenic viruses. Annu Rev Microbiol 1:85–100

Kurath G, Palukaitis P (1989) RNA sequence heterogeneity in natural populations of three satellite RNAs of cucumber mosaic virus. Virology 173:231–240

Li H, Roossinck MJ (2004) Genetic bottlenecks reduce population variation in an experimental RNA virus population. J Virol 78:10582–10587

Malmstrom CM, Stoner CJ, Brandenburg S, Newton LA (2006) Virus infection and grazing exert counteracting influences on survivorship of native bunchgrass seedling competing with invasive exotics. J Ecol 94:264–275

Malpica JM, Fraile A, Moreno I, Obiés CI, Drake JW, García-Arenal F (2002) The rate and character of spontaneous mutation in an RNA virus. Genetics 162:1505–1511

Malpica JM, Sacristán S, Fraile A, García-Arenal F (2006) Association and host selectivity in multi-host pathogens. PLoS ONE 1:e41

Márquez LM, Redman RS, Rodríguez RJ, Roossinck MJ (2007) A virus in a fungus in a plant-three way symbiosis required for thermal tolerance. Science (in press)
Martin DP, van der Walt E, Posada D, Rybicki EP (2005) The evolutionary value of recombination is constrained by genome modularity. PLoS Genet 1:e51
Maskell LC, Raybould AF, Cooper JI, Edwards M-L, Gray AJ (1999) Effects of turnip mosaic and turnip yellow mosaic virus on the survival, growth and reproduction of wild cabbage (*Brassica oleracea*). Ann Appl Biol 135:401–407
Mastari J, Lapierre H, Dessens JT (1998) Asymmetrical distribution of barley yellow dwarf virus PAV variants between host plant species. Phytopathology 88:818–821
Moreno I, Malpica JM, Rodríguez-Cerezo E, García-Arenal F (1997) A mutation in tomato aspermy cucumovirus that abolishes cell-to-cell movement is maintained to high levels in the viral RNA population by complementation. J Virol 71:9157–9162
Moury B (2004) Differential selection of genes of cucumber mosaic virus subgroups. Mol Biol Evol 21:1602–1611
Pelham J, Fletcher JT, Hawkins JH (1970) The establishment of a new strain of tobacco mosaic virus resulting from the use of resistant varieties of tomato. Ann Appl Biol 65:293–297
Posada D, Crandall KA (2001) Evaluation of methods for detecting recombination from DNA sequences: computer simulations. Proc Natl Acad Sci USA 98:13757–13762
Power AG, Mitchell CE (2004) Pathogen spillover in disease epidemics. Am Nat 164:S79–S89
Reddy DVR, Black LM (1977) Isolation and replication of mutant populations of wound tumor virions lacking certain genome segments. Virology 80:336–346
Rodríguez-Cerezo E, García-Arenal F (1989) Genetic heterogeneity of the RNA genome population of the plant virus U5-TMV. Virology 170:418–423
Rodríguez-Cerezo E, Elena SF, Moya A, García-Arenal F (1991) High genetic stability in natural populations of the plant RNA virus tobacco mild green mosaic virus. J Mol Evol 32:328–332
Sacristán S, Malpica JM, Fraile A, García-Arenal F (2003) Estimation of population bottlenecks during systemic movement of *Tobacco mosaic virus* in tobacco plants. J Virol 77:9906–9911
Sacristán S, Fraile A, Malpica JM, García-Arenal F (2005) An analysis of host adaptation and its relationship with virulence in *Cucumber mosaic virus*. Phytopathology 95:827–833
Sanjuán R, Moya A, Elena SF (2004a) The distribution of fitness effects caused by single-nucleotide substitutions in an RNA virus. Proc Natl Acad Sci USA 101:8396–8401
Sanjuán R, Moya A, Elena SF (2004b) The contribution of epistasis to the architecture of fitness in an RNA virus. Proc Natl Acad Sci USA 101:15376–15379
Sanz AI, Fraile A, García-Arenal F, Zhou X, Robinson DJ, Khalid S, Butt T, Harrison BD (2000) Multiple infection, recombination and genome relationships among begomovirus isolates found in cotton and other plants in Pakistan. J Gen Virol 81:1839–1849
Schneider WL, Roossinck MJ (2000) Evolutionary related Sindbis-like plant viruses maintain different levels of population diversity in a common host. J Virol 74:3130–3134
Schneider WL, Roossinck MJ (2001) Genetic diversity in RNA virus quasispecies is controlled by host-virus interactions. J Virol 75:6566–6571
Simón B, Cenis JL, Beitia F, Khalid S, Moreno IM, Fraile A, García-Arenal F (2003) Genetic structure of field populations of Begomoviruses and of their vector *Bemisia tabaci* in Pakistan. Phytopathology 93:1422–1429
Skotnicki ML, Mackenzie AM, Ding SW, Mo JQ, Gibbs AJ (1993) RNA hybrid mismatch polymorphisms in Australian populations of turnip yellow mosaic tymovirus. Arch Virol 132:83–99
Skotnicki ML, Mackenzie AM, Gibbs AJ (1996) Genetic variation in populations of kennedya yellow mosaic tymovirus. Arch Virol 141:99–110
Stewart AD, Logson JM Jr, Kelley SE (2005) An empirical study of the evolution of virulence under both horizontal and vertical transmission. Evolution 59:730–739
Tan Z, Wada Y, Chen J, Ohshima K (2004) Inter- and intralineage recombinants are common in natural populations of *Turnip mosaic virus*. J Gen Virol 85:2683–2696
Tepfer M (2002) Risk assessment of virus-resistant transgenic plants. Annu Rev Phytopathol 40:467–91

Tomitaka Y, Ohshima K (2006) A phylogeographical study of the *Turnip mosaic virus* population in East Asia reveals an "emergent" lineage in Japan. Mol Ecol 15:4437–4457
Van Regenmortel MHV (1975) Antigenic relationships between strains of tobacco mosaic virus. Virology 64:415–420
White PS, Morales FJ, Roossinck MJ (1995) Interspecific reassortment in the evolution of a cucumovirus. Virology 207:334–337
Wilke CO (2005) Quasispecies theory in the context of population genetics. BMC Evol Biol 5:44
Woolhouse MEJ, Taylor LH, Haydon DT (2001) Population biology of multihost pathogens. Science 292:1109–1112
Yang Z, Nielsen R, Goldman N, Krabbe Pedersen A-M (2000) Codon-substitution models for heterogeneous selection pressure at amino acid sites. Genetics 155:431–449
Yarwood CE (1979) Host passage effects with plant viruses. Adv Virus Res 24:169–190
Zanotto PM, Gibbs MJ, Gould EA, Holmes EC (1996) A reevaluation of the higher taxonomy of viruses based on RNA polymerases. J Virol 70:6083–6096

Chapter 2
Community Ecology of Plant Viruses

Alison G. Power

Abstract The ecology of plant viruses has been little studied, particularly in natural ecosystems where symptoms often go unnoticed. Interactions between a plant virus and its host plant are embedded in a broader community of species, many of which can influence the dynamics of both virus and plant. Plant viruses tend to have a limited number of effective vectors, and vector population dynamics, host preference, and movement have a strong influence on virus ecology. Vector dynamics and behavior are strongly influenced by the diversity and structure of the plant community. Ecological interactions between viruses within a host, such as cross-protection, mutual suppression, or competitive exclusion, also impact virus populations. The potential for linked disease and community dynamics is illustrated by experiments that reveal community-shaping apparent competition between plants, resulting from the spillover of plant viruses from reservoir hosts to less susceptible hosts. Moreover, virus suppression of otherwise dominant plant species can have significant consequences in plant communities. Recent studies on the role of viruses in plant invasions suggest that landscape-level changes have resulted from the interactions of viruses and hosts within a community context that includes competing plant species, reservoir hosts, herbivores, and microbial symbionts of plants. As ecological research on plant viruses in natural ecosystems increases, we are likely to see more examples of strong impacts of viruses on the structure of plant communities.

Alison G. Power
350 Caldwell Hall, Cornell University, Ithaca, NY 14853, USA
agp4@cornell.edu

M.J. Roossinck (ed.), *Plant Virus Evolution.*

2.1 Introduction

Although the genetics and pathogenicity of plant viruses have been well studied in crop hosts, their diversity and ecology has received less attention, particularly in natural ecosystems where symptoms are often less noticeable (Wren et al. 2006). There has been a recent flurry of studies on the prevalence and impact of plant viruses in wild host plants, partly as a result of ecological risk assessment for transgenic virus resistance in major crops (reviewed in Cooper and Jones 2006). While the strength of the interactions between viruses and hosts can vary, these studies confirm the potential of viruses to influence the fitness of host populations (Friess and Maillet 1996; Funayama et al. 1997; Maskell et al. 1999; Power and Mitchell 2004; Malmstrom et al. 2005a, 2006).

Despite this increase in recent research, we still know very little about how plant viruses interact with plant communities. Our understanding of virus epidemiology comes largely from agricultural systems, where there is little potential for long-term feedbacks from pathogens to host population dynamics because the host population is directly controlled by humans (Mitchell and Power 2006). Many studies have shown that the diversity and structure of cropping systems can influence the spread of plant pathogens, including viruses (Power 1987, 1991; Mundt 2002). These provide the foundation for understanding virus ecology in plant communities.

In natural ecosystems, plant pathogens have been shown to drive plant population dynamics, mediate plant competition, modify plant community structure, maintain plant species diversity, and cause rapid evolution in plant populations (Gilbert 2002; Mitchell and Power 2006). However, most of the pathogens that have been studied have been fungal pathogens, whereas there is much less information about the role of plant viruses. Similarly, there is increasing evidence that the spread of plant pathogens is influenced by the density and genetic structure of host populations, but again, most of these studies have focused on fungal pathogens (Mitchell and Power 2006). The potential for reciprocal effects between viruses and the structure of plant communities is clear, but there are few data available that allow us to quantify these effects.

Among the many factors that shape the ecology and dynamics of plant viruses, two major characteristics of plants are critical: their relative immobility and their lack of highly specific immune systems. The first characteristic leads directly to the heavy dependence of plant viruses on effective transmission mechanisms. While some viruses are transmitted vertically from parent to offspring in seed, most plant viruses depend on horizontal transmission by vectors. The virus's ability to move from an immobile infected host to an immobile healthy host is key (Ng and Falk 2006). The second characteristic suggests that plants are likely to employ generalized defense strategies against viruses and that viruses will be under selection pressure to overcome these general defenses. Both of these factors shape the pattern and outcome of interactions between plant viruses and plant communities.

In this chapter I address three aspects of the community ecology of plant viruses. First, I summarize patterns of host range and vector range among plant viruses, based on an analysis of 910 unique species of plant viruses (Power and Flecker 2003, 2007). These patterns highlight the importance of vector relations in determining the distribution of viruses in plant communities. Second, I describe several studies that examine how virus communities are shaped by within-host processes and transmission by vectors. I argue that between-host processes are often more important than within-host processes in shaping virus epidemiology. Finally, I discuss recent evidence demonstrating the potential for strong impacts of viruses on the structure of natural plant communities, using the widespread barley and cereal dwarf viruses as a case study. Throughout, I emphasize the important role of vectors in shaping community structure through virus transmission.

2.2 Patterns of Host and Vector Specialization

Plant viruses are likely to be host generalists and vector specialists; that is, most viruses have a large host range and a very narrow range of vectors (Power and Flecker 2003, 2007; Fig. 2.1). Most plant viruses are transmitted by vectors, and of these, approximately 60% are transmitted by a single species of vector (Power and Flecker 2003). In contrast, less than 10% of plant viruses infect a single species of host. This pattern becomes even more striking when analyzing it at the genus or family level, where some viruses infect dozens of plant families with vectors from a single genus. Virus host range appears to be driven largely by the host range of the vector, suggesting that virus–vector interactions, rather than virus–host interactions, are controlling disease spread for many plant viruses (Power and Flecker 2003).

DNA viruses seem particularly likely to have low vector diversity. Of the vector-transmitted DNA viruses in the VIDE database on plant viruses (Brunt et al. 1996), 83.8% were transmitted by a single species of vector, whereas 49.1% of RNA viruses had one vector species (Power and Flecker 2007). RNA viruses may be more likely to adapt to multiple vectors owing to their high mutation rates. Mutation rates may be as much as 300 times higher in RNA viruses than in DNA viruses, and higher mutation rates should lead to greater genetic variation in the population (Woolhouse et al. 2001). Despite these high mutation rates of RNA viruses, genetic bottlenecks that occur during vector transmission have been demonstrated to constrain variation, since not all genetic variants are included in a single transmission event (Ali et al. 2006). Interestingly, in experiments with *Cucumber mosaic virus*, genetic variation was not reduced as the aphid acquired the virus from the host plant, but was reduced during the process of transmitting the virus to a new host (Ali et al. 2006).

Woolhouse et al. (2005) argued that RNA viruses are more likely than DNA viruses to jump to new host species. In contrast, we found no significant difference in the host range of DNA versus RNA plant viruses ($P>0.10$). Other studies of virus–host interactions have shown that mutation rates may be correlated with host range within a virus (Schneider and Roossinck 2001). Our analysis detected a different relationship between host range and genome structure, in that single-stranded RNA and DNA viruses had

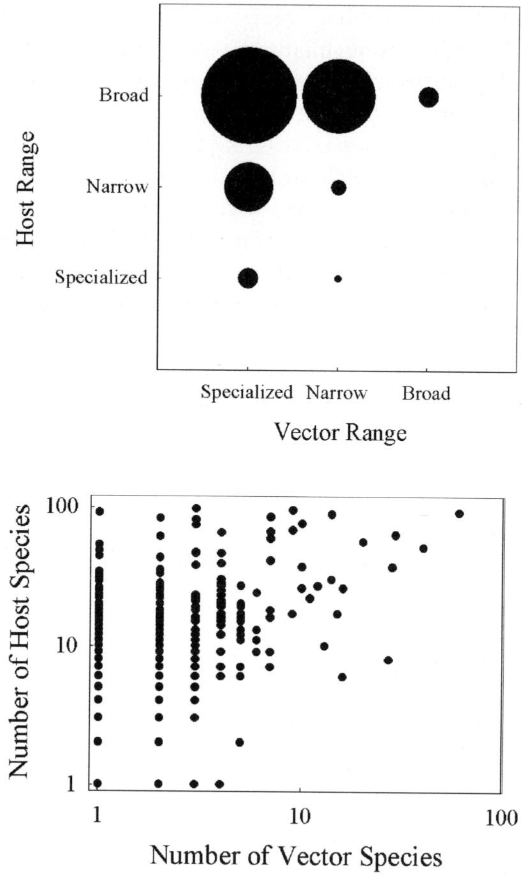

Fig. 2.1 Number of hosts and number of vectors for vector-transmitted plant viruses (n=474). **a** Proportion of vector-transmitted plant viruses that fall into nine general categories of host and vector range. **b** Number of host species versus number of vector species for each virus; each point represents one virus species, but some points overlap. (From Power and Flecker 2003)

significantly more hosts than double-stranded viruses (P<0.001), suggesting that single-stranded viruses may adapt more readily to new hosts. However, vector associations explained a much greater proportion of the variance in host range (R^2=0.31) than did genome characteristics (R^2=0.01). Both the number of vector species and the type of vector (insect, fungus, mite or nematode) were major determinants of host range. Although genetic bottlenecks have been shown to occur during within-plant systemic movement from an infection site (Li and Roossinck 2004; Moury et al. 2006), the overall pattern described above suggests that this reduced variation may not strongly limit viral host range.

These patterns may imply that plant-to-plant movement (i.e., attaining access to new hosts) is at least as important as overcoming plant defenses in determining rates of virus spread. Many host antiviral mechanisms such as gene silencing are highly generalized and operate against a large range of viruses. Viral invasion can induce gene silencing and provide cross-protection against secondary virus infection; however, suppression of gene silencing is a general strategy used by a broad range of DNA and RNA plant viruses (Voinnet et al. 1999; Voinnet 2005). In fact, most known viral silencing suppressor mechanisms suppress RNA silencing in both animal and plant cells, regardless of the host origin of the virus (Qu and Morris 2005).

Disease dynamics can be dramatically influenced by virus adaptation to a new vector species. For example, the occurrence and range of begomoviruses increased dramatically in recent years as a result of the introduction of the Old World B-biotype *Bemesia tabaci* whitefly to the Americas. B-biotype *B. tabaci* have an unusually broad host range and transmit begomoviruses among host plants, both cultivated and wild, that did not previously share insect vectors (Brown et al. 1995). The introduction of this vector to new geographic areas provided the opportunity for pre-existing viruses to be transmitted to a variety of new hosts. This broad vector host range may also increase the frequency of mixed infections of begomoviruses, leading to the emergence of new viruses that result from recombination among strains or species (Seal et al. 2006). Moreover, the high genetic variation in geminiviruses suggest that they should be considered quasispecies (Roossinck 1997), so vector transmission is likely to cause genetic bottlenecks as well as exerting selection on the virus population.

Although our understanding of genetic controls on transmission is still limited, molecular analyses have repeatedly suggested that areas of the viral genome responsible for regulating the specificity of insect transmission are more highly conserved than those for host infection (Power 2000). This conservation has been shown for both RNA viruses and DNA viruses. Even viruses that are simply carried on the mouthparts of vectors depend on complex interactions between viral proteins and vector-associated compounds (Gray and Banerjee 1999). Nault (1997) has argued that the mechanism of transmission is a stable evolutionary trait for virus genera, and it is clear that the evolution of plant viruses is constrained by the need to retain specific interactions with their vectors. In several cases, expansion of the host range of insect vectors has been shown to increase the host range of the viruses that these vectors transmit (Harrison and Robinson 1999; Goldbach and Peters 1994), implying that virus distribution depends crucially on the specificity of virus–vector relations. Patterns of virus prevalence are largely determined by the host preference, feeding behavior, aggregation behavior and movement behavior of vectors, as well as vector population dynamics.

2.3 Virus Interactions Within and Among Hosts

Most plant species can be infected by more than one virus, and viruses often occur in mixed infections. From laboratory studies, we know that mixed infections may lead to a variety of competitive or facilitative interactions within the host, including synergism, cross-protection, mutual suppression, replacement, and helper-dependence

(Rochow 1972; Power 1992; Hammond 1999). To date, we have limited understanding of the importance of these interactions and how they might influence virus epidemiology (Jeger et al. 2006) or modify virus impacts on hosts in natural ecosystems. Rates of mixed infections may be determined by these within-host interactions that affect virus replication and persistence or by population-level processes such as increased mortality due to synergy, the presence of shared vectors, or modifications of transmission processes due to coinfection.

There are a number of examples of the apparent displacement of one virus by another but the causal mechanism of displacement is often obscure. In some cases, such as the rise of resistance-breaking strains, interactions with the host are clearly involved (Harrison 2002). In other cases, a more complex set of interactions may be responsible. The barley and cereal yellow dwarf viruses are phloem-limited single-stranded RNA luteoviruses that are obligately transmitted in a persistent, circulative manner by several species of grass-feeding aphids to cultivated and wild grasses (Miller and Rasochova 1997). A 20–year shift in the distribution of two of these viruses, *Barley yellow dwarf virus–PAV* (BYDV-PAV) and *Barley yellow dwarf virus–MAV* (BYDV-MAV), was documented by Rochow (1979). A retrospective analysis of the displacement suggested that asymmetric competition and cross-protection between PAV and MAV may have played a role in the displacement, but differences between viruses in the number of vector species and the relative efficiency of transmission by these vectors were probably more important influences on prevalence (Power 1996).

In recent field studies, we investigated the naturally occurring distribution of five coexisting barley and cereal yellow dwarf viruses in *Elymus glaucus*, a native California grass (E. Seabloom, A. Power, and E. Borer, unpublished data). We found that coinfection rates were much higher than expected, suggesting that cross-protection was not exerting strong effects on virus populations within hosts. Moreover, we found that the co-occurrence of viruses was not correlated with their identification as a barley yellow dwarf virus (group 1) or a cereal yellow dwarf virus (group II), which would be expected if cross-protection or synergistic mortality were operating within or between the two groups of viruses (Miller and Rasochova 1997). Instead, the covariance of individual strains across a 700-km latitudinal gradient and over 4 years was highest for strains that shared an aphid vector species. These results support the hypothesis that the vector community drives virus dynamics at large spatial and temporal scales.

2.4 Virus Spread in a Community Context

Most plant viruses have multiple hosts, yet most theoretical and empirical studies of virus dynamics are restricted to single host species. Because host species vary in critical epidemiological traits such as resistance, tolerance, and attractiveness to vectors, transmission rates within and between different host species are variable and asymmetric (Daszak et al. 2000; LoGiudice et al. 2003; Woolhouse et al. 2001).

In some host populations, infection may be largely determined by the host community as a whole. In cases where virus transmission within a host population is insufficient to maintain virus populations, transmission from a reservoir species may sustain virus spread. The virus reaches high prevalence in the reservoir host, then spills over into a less suitable host, a process called "the spillover effect" or "pathogen spillover" (Daszak et al. 2000). As a result of pathogen spillover, understanding virus ecology requires considering the community context in which the virus and its host are embedded.

To investigate the role of pathogen spillover in the ecology of plant viruses, we carried out a series of field experiments with BYDV-PAV, which infects a broad range of grass hosts. In experiments with constructed communities of wild annual grasses, the presence of a reservoir species, *Avena fatua* (wild oats), greatly increased the prevalence of BYDV-PAV in several other species (Power and Mitchell 2004). When *A. fatua* was added to a grass community consisting of *Lolium multiflorum*, *Setaria lutescens*, and *Digitaria sanguinalis*, virus spillover from *A. fatua* led to higher virus prevalence across the other three less heavily infected hosts, approximately doubling virus prevalence in all of the other species (Fig. 2.2).

As a host, *A. fatua* is highly susceptible to BYDV, which quickly builds up high titers in the host. In addition, *A. fatua* is very attractive to several species of aphids (Malmstrom et al. 2005b; Lowry 2007). In preference trials, two important species of aphid vectors, *Rhopalosiphum padi* and *Sitobion avenae*, both strongly preferred *A. fatua* to *L. multiflorum*, *S. lutescens*, and *D. sanguinalis*, the three other hosts in this system (Lowry 2007). Both aphid species also had highest fecundity on *A. fatua*, suggesting that this host can amplify vector populations as well as virus populations.

In these grass communities, virus spillover from *A. fatua* resulted in apparent competition between *A. fatua* and both *L. multiflorum* and *D. sanguinalis* (Power and Mitchell 2004). *A. fatua* is a strong competitor with other grasses, and in this system it has a strong negative competitive effect on *S. lutescens*. When *A. fatua* was added to communities without BYDV, the presence of *A. fatua* increased the aboveground biomass of *L multiflorum* and *D. sanguinalis* by suppressing the dominant *S. lutescens* (Fig. 2.3). That is, *L. multiflorum* and *D. sanguinalis* benefited

Fig. 2.2 Virus prevalence in *Digitaria sanguinalis* (*black bars*), *Lolium multiflorum* (*white bars*), and *Setaria lutescens* (*gray bars*) in experimental communities planted with *Avena fatua* or lacking *A. fatua*. The presence of *A. fatua* increased virus prevalence across the other three species, demonstrating pathogen spillover. (From Power and Mitchell 2004)

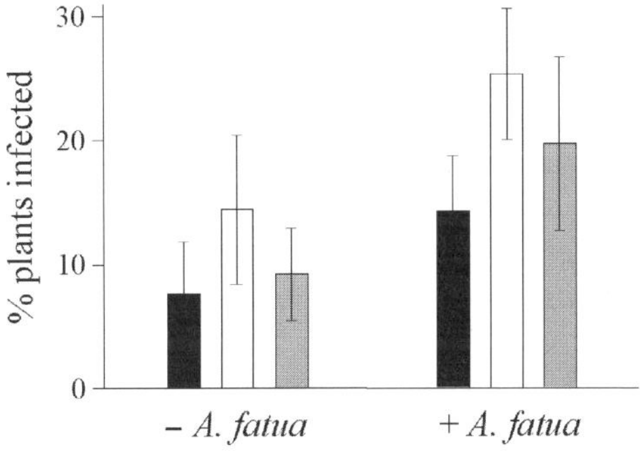

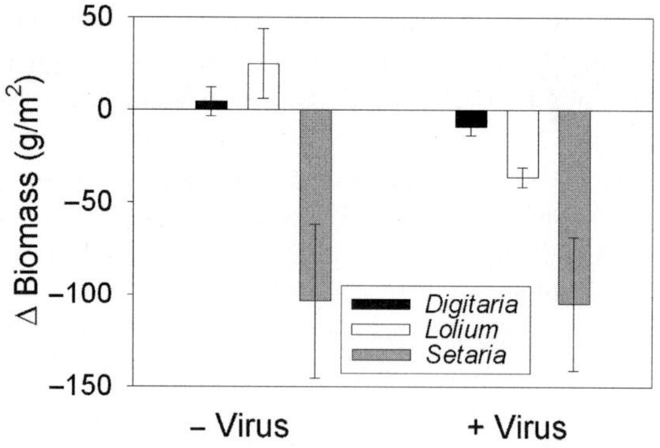

Fig. 2.3 The effect of adding *A. fatua* on aboveground vegetative biomass of *D. sanguinalis* (*black bars*), *L. multiflorum* (*white bars*), and *S. lutescens* (*gray bars*) in experimental communities. Communities were either inoculated with the virus (+ *virus*) or mock inoculated using nonviruliferous aphids (– *virus*). Adding *A. fatua* decreased *D. sanguinalis* and *L. multiflorum* biomass only in virus-inoculated communities, demonstrating virus-mediated apparent competition. (From Power and Mitchell 2004)

more from *A. fatua*'s strong negative competitive effect on *S. lutescens* than they were harmed by direct competition with *A. fatua*. When BYDV was present, however, adding *A. fatua* decreased the biomass of these two species. This appeared to be the outcome of direct competition, but in fact resulted from pathogen spillover from *A. fatua*. In this case, the negative effects of *A. fatua* on *L. multiflorum* and *D. sanguinalis* were mediated by the presence of the shared pathogen, resulting in apparent competition. In contrast, apparent competition had little or no effect on *S. lutescens*, despite increased virus prevalence as a result of spillover from *A. fatua*. This suggested that *S. lutescens* was more tolerant of BYDV infection than the other two host species (Fig. 2.3). Tolerance may be a key determinant of the outcome of pathogen-mediated apparent competition. Overall, these results provided experimental evidence for the ecological importance of virus spillover and apparent competition in plant communities.

2.5 Viruses and Plant Invasions

The emerging story of how barley and cereal yellow dwarf viruses may have influenced the invasion of Californian grasslands by exotic annual grasses is a dramatic example of complex ecological interactions involving viruses and grasses (Malmstrom et al. 2005a; Borer et al. 2007). Populations of many native perennial bunchgrasses have declined dramatically over the past century, largely replaced by these introduced annual grasses. While this decline has often been attributed to competition and disturbance, research has shown that, paradoxically, the annual grasses are typically inferior competitors (Seabloom et al. 2003; Corbin and

D'Antonia 2004). Recent studies have suggested that BYDV may have been an additional factor contributing to the decline of the native perennial grasses. Malmstrom et al. (2005b) found dramatically increased prevalence of barley yellow dwarf viruses in a native Californian perennial bunchgrass, *E. glaucus*, growing with the introduced annual *A. fatua*, compared with bunchgrasses growing in pure stands. Since virus infection reduces biomass and seed production of *E. glaucus* and other native perennials (Malmstrom et al. 2005a), these results imply that apparent competition may be at least partially responsible for the negative impact of exotic annuals like *A. fatua* on native bunchgrasses in Californian grasslands. Additional experiments indicated that the direct effects of BYDV on bunchgrass survivorship were enhanced by competition with annual grasses (Malmstrom et al. 2006). A dynamic model with field-estimated parameters suggested that the presence of BYDV in the system could reverse the normal competitive dominance of native perennials, allowing exotic annuals to invade and dominate (Borer et al. 2007). Taken together, these studies imply that a group of plant viruses may have played an important role in a series of plant invasions that transformed a landscape.

Other studies have addressed the role of viruses in plant invasions more generally. In a test of the "enemy release hypothesis," Mitchell and Power (2003) showed that 24% fewer plant viruses and 84% fewer aboveground fungal pathogens (rusts, smuts, and powdery mildew) infected plants in their naturalized range than in their native range. Invasive plants that escaped the largest proportion of their native pathogens were most likely to be considered harmful invaders in both agricultural and natural ecosystems. These patterns suggest that plants escape many pathogens during the invasion process. It is likely that plants escape viruses less easily than fungi because viruses tend to have broader host ranges, and they are more likely to be systemic and asymptomatic in their hosts (Mitchell and Power 2003). The impacts of viruses on invasive plants, as well as other hosts in natural ecosystems, are likely to be modulated by other plant interactions, such as competition with other plants, mutualisms with other microorganisms, or herbivory (Mitchell et al. 2006).

2.6 Summary

Interactions between a plant virus and its host plant are embedded in a broader community of species, many of which can influence the dynamics of both virus and plant. This is especially true for viruses which are host generalists, which includes the majority of plant viruses. Interactions between viruses and plants will be modified by parasitism, competition, mutualism, and herbivory – all the typical interactions that occur in communities (Mitchell and Power 2006). Here we have focused on the important role of vectors in shaping the dynamics of virus populations. Plant viruses tend to have a limited number of effective vectors, and vector population dynamics, host preference, and movement have a strong influence on virus ecology. Ecological interactions between viruses within a host, such as cross-protection, mutual suppression, or competitive exclusion, may also impact virus populations.

However, empirical studies suggest that relations with vectors are often of overriding importance in determining the distribution of viruses.

Research on the impact of viruses on plant communities is in its infancy, but it is already apparent that there may be profound ecological repercussions in natural ecosystems. The potential for linked disease and community dynamics is illustrated by the community-shaping apparent competition that results from generalist virus spillover from reservoir species to less susceptible species. It is clear that virus suppression of otherwise dominant host species can have significant consequences in plant communities. Recent research on the role of viruses in plant invasions suggest substantial, landscape-level changes resulting from the interactions of viruses and hosts within a community context that includes competing plant species, reservoir hosts, herbivores, and microbial symbionts of plants. Examples such as the ones described in this chapter are becoming more common as ecological research on viruses increases.

Acknowledgements This chapter is based upon research supported by Cornell University and the National Science Foundation (grants no. 0444639 and 0525669).

References

Ali A et al (2006) Analysis of genetic bottlenecks during horizontal transmission of Cucumber mosaic virus. J Virol 80:8345–8350

Borer ET, Hosseini PR, Seabloom EW, Dobson AP (2007) Pathogen-induced reversal of native dominance in a grassland community. Proc Natl Acad Sci USA 104:5473–5478

Brown JK, Frolich DR, Rosell RC (1995) The sweetpotato or silverleaf whiteflies: biotypes of *Bemisia tabaci* or a species complex? Annu Rev Entomol 40:511–534

Brunt AA, Crabtree K, Dallwitz MJ, Gibbs AJ, Watson L, Zurcher EJ (1996) Plant viruses online: descriptions and lists from the VIDE database. http://biology.anu.edu.au/Groups/MES/vide/

Cooper I, Jones RAC (2006) Wild plants and viruses: under-investigated ecosystems. Adv Virus Res 67:1–47

Corbin JD, D'Antonio CM (2004) Competition between native perennial and exotic annual grasses: Implications for an historical invasion. Ecology 85:1273–1283

Daszak P, Cunningham AA, Hyatt AD (2000) Emerging infectious diseases of wildlife – threats to biodiversity and human health. Science 287:443–449

Friess N, Maillet J (1996) Influence of cucumber mosaic virus infection on the intraspecific competitive ability and fitness of purslane (*Portulaca oleracea*). New Phytol 132:103–111

Funayama S, Hikosaka K, Yahara T (1997) Effects of virus infection and growth irradiance on fitness components and photosynthetic properties of Eupatorium makinoi (Compositae). Am J Bot 84:823–829

Gilbert GS (2002) Evolutionary ecology of plant diseases in natural ecosystems. Annu Rev Phytopathol 40:13–43

Goldbach R, Peters D (1994) Possible causes of the emergence of tospovirus diseases. Semin Virol 5:113–120

Gray SM, Banerjee N (1999) Mechanisms of arthropod transmission of plant and animal viruses. Microbiol Mol Biol Rev 63:128

Hammond J, Lecoq H, Raccah B (1999) Epidemiological risks from mixed virus infections and transgenic plants expressing viral genes. Adv Virus Res 54:189–314

Harrison BD (2002) Virus variation in relation to resistance-breaking in plants. Euphytica 124: 181–192
Harrison BD, Robinson DJ (1999) Natural genomic and antigenic variation in whitefly-transmitted geminiviruses (Begomoviruses). Annu Rev Phytopathol 37:369–398
Jeger MJ, Seal SE, Van den Bosch F (2006) Evolutionary epidemiology of plant virus disease. Adv Virus Res 67:163–203
Li HY, Roossinck MJ (2004) Genetic bottlenecks reduce population variation in an experimental RNA virus population. J Virol 78:10582–10587
LoGiudice K, Ostfeld RS, Schmidt KA, Keesing F (2003) The ecology of infectious disease: effects of host diversity and community composition on Lyme disease risk. Proc Natl Acad Sci USA 100:567–571
Lowry E (2007) The role of aphid host preference in barley yellow dwarf virus epidemiology. Undergraduate honors thesis, Cornell University, Ithaca, p 26
Malmstrom CM, McCullough AJ, Johnson HA, Newton LA, Borer ET (2005a) Invasive annual grasses indirectly increase virus incidence in California native perennial bunchgrasses. Oecologia 145:153–164
Malmstrom CM, Hughes CC, Newton LA, Stoner CJ (2005b) Virus infection in remnant native bunchgrasses from invaded California grasslands. New Phytol 168:217–230
Malmstrom CM, Stoner CJ, Brandenburg S, Newton LA (2006) Virus infection and grazing exert counteracting influences on survivorship of native bunchgrass seedlings competing with invasive exotics. J Ecol 94:264–275
Maskell LC, Raybould AF, Cooper JI, Edwards ML, Gray AJ (1999) Effects of turnip mosaic virus and turnip yellow mosaic virus on the survival, growth and reproduction of wild cabbage (*Brassica oleracea*). Ann Appl Biol 135:401–407
Miller WA, Rasochova L (1997) Barley yellow dwarf viruses. Annu Rev Phytopathol 35:167–190
Mitchell CE, Power AG (2003) Release of invasive plants from fungal and viral pathogens. Nature 421:625–627
Mitchell CE, Power AG (2006) Disease dynamics in plant communities. In: Collinge SK, Ray C (eds) Community structure and pathogen dynamics. Oxford University Press, Oxford, pp 58–72
Mitchell CE et al (2006) Biotic interactions and plant invasions. Ecol Lett 9:726–740
Moury B, Desbiez C, Jacquernond M, Lecoq H (2006) Genetic diversity of plant virus populations: Towards hypothesis testing in molecular epidemiology. Adv Virus Res 67:49–87
Mundt CC (2002) Use of multiline cultivars and cultivar mixtures for disease management. Annu Rev Phytopathol 40:381–410
Nault LR (1997) Arthropod transmission of plant viruses: a new synthesis. Ann Entomol Soci Am 90:521–541
Ng JCK, Falk BW (2006) Virus-vector interactions mediating nonpersistent and semipersistent transmission of plant viruses. Annu Rev Phytopathol 44:183–212
Power AG (1987) Plant community diversity, herbivore movement, and an insect-transmitted disease of maize. Ecology 68:1658–1669
Power AG (1991) Virus spread and vector dynamics in genetically diverse plant populations. Ecology 72:232–241
Power AG (1992) Patterns of virulence and benevolence in insect-borne pathogens of plants. Crit Rev Plant Sci 11:351–372
Power AG (1996) Competition between viruses in a complex plant-pathogen. Ecology 77:1004–1010
Power AG (2000) Insect transmission of plant viruses: a constraint on virus variability. Curr Opin Plant Biol 3:336–340
Power AG, Flecker AS (2003) Virus specificity in disease systems: Are species redundant? In: Levin SA, Kareiva P (eds) The importance of species: perspectives on expendability and triage. Princeton University Press, Princeton, pp 330–346

Power AG, Flecker AS (2007) The role of vector diversity in disease dynamics. In: Ostfeld RS, Keesing F, Eviner VT (eds) Infectious disease ecology: the effects of ecosystems on disease and of disease on ecosystems. Princeton University Press, Princeton
Power AG, Mitchell CE (2004) Pathogen spillover in disease epidemics. Am Nat 164:S79–S89
Qu F, Morris TJ (2005) Suppressors of RNA silencing encoded by plant viruses and their role in viral infections. FEBS Lett 579:5958–5964
Rochow WF (1972) Role of mixed infections in transmission of plant viruses by aphids. Annu Rev Phytopathol 10:101–124
Rochow WF (1979) Field variants of barley yellow dwarf virus – detection and fluctuation during 20 years. Phytopathology 69:655–660
Roossinck MJ (1997) Mechanisms of plant virus evolution. Annu Rev Phytopathol 35:191–209
Schneider WL, Roossinck MJ (2001) Genetic diversity in RNA virus quasispecies is controlled by host-virus interactions. J Virol 75:6566–6571
Seabloom EW, Harpole WS, Reichman OJ, Tilman D (2003) Invasion, competitive dominance, and resource use by exotic and native California grassland species. Proc Natl Acad Sci USA 100:13384–13389
Seal SE, Jeger MJ, Van den Bosch F (2006) Begomovirus evolution and disease management. Adv Virus Res 67:297–316
Voinnet O (2005) Induction and suppression of RNA silencing: insights from viral infections. Nat Rev Genet 6:206–220
Voinnet O, Pinto YM, Baulcombe DC (1999) Suppression of gene silencing: a general strategy used by diverse DNA and RNA viruses of plants. Proc Natl Acad Sci USA 96:14147–14152
Woolhouse MEJ, Taylor LH, Haydon DT (2001) Population biology of multihost pathogens. Science 292:1109–1111
Woolhouse MEJ, Haydon DT, Antia R (2005) Emerging pathogens: the epidemiology and evolution of species jumps. Trends Ecol Evol 20:238–244
Wren JD, Roossinck MJ, Nelson RS, Scheets K, Palmer MW, Melcher U (2006) Plant virus biodiversity and ecology. PLoS Biol 4:314–315

Chapter 3
Emerging Plant Viruses: a Diversity of Mechanisms and Opportunities

Maria R. Rojas(✉) and Robert L. Gilbertson

Abstract Although emerging plant viruses receive much less publicity than their animal- or human-infecting cousins, they pose a serious threat to worldwide

---

Maria R. Rojas
Department of Plant Pathology, University of California, Davis, CA, USA
mjrojas@ucdavis.edu

M.J. Roossinck (ed.), *Plant Virus Evolution*.

agricultural production. These viruses can be new (i.e., not previously known) or already known; however, they share the common characteristic of occupying and spreading within new niches. Factors driving the emergence of plant viruses include genetic variability in the virus, changes in agricultural practices, increases in the population and/or distribution of insect vectors and long-distance transport of plant materials. In recent years, individual as well as entire groups of viruses have emerged, and this has involved a variety of mechanism(s), depending on the virus and the environment. Here, we will discuss some of these viruses, and highlight the mechanisms that have mediated their emergence. Special emphasis is placed upon the whitefly-transmitted geminiviruses (begomoviruses) and the thrips-transmitted tosposviruses, which have emerged as major threats to crop production throughout the world. Other examples include the recent emergence of novel viruslike agents, the acquisition and role of satellite DNA or RNA molecules in emergence of plant viruses, and cases where emerging viruses have had only a transient impact. It seems clear that global movement of plant materials, expansion of agriculture and large-scale monoculture will continue to favor emergence of plant viruses. However, improved diagnostics should allow for rapid identification of emerging viruses and better understanding of viral biology. This information can be used in the development of effective management strategies, which will hopefully minimize impact on agricultural production.

3.1 Introduction

Emerging viruses pose a major threat to animal, human and plant health. Some of these viruses have already greatly impacted human health, whereas others have caused catastrophic losses to crop and animal production. There are many definitions of what constitutes an emerging virus. We consider an emerging virus to be one that has recently changed or appeared to occupy and spread within a new niche. Emerging viruses can be new, i.e., not previously known; however, they are often known viruses that have become more apparent owing to changes in the environment/ecosystem and/or generation of a new variant, thereby providing the virus with an opportunity to expand into new niches. In the case of plant viruses, changes in agricultural practices and long-distance transport of plant materials are key factors mediating virus emergence. Some emerging viruses gain considerable public recognition and attention because of actual or potential health or economic losses, e.g., development of animal or human diseases or economic losses due to disease epidemics in crop plants. In other cases, the emergence of a virus or group of viruses may not result in catastrophic disease or economic losses.

Viruses that have emerged as animal or human pathogens are usually much more highly publicized than are emergent viruses of plants. However, the rate of emergence of plant viruses does not seem to be any less than that for human and other animal viruses. Furthermore, some common mechanisms underlie the emergence of animal and plant viruses, irrespective of the nature of the host. For example, an increasingly important mechanism is inadvertent long-distance transport of viruses, which is mediated by increased global movement of plants and people. Another common mechanism is the capacity of viruses to jump from reservoir hosts into new hosts; these viruses can cause serious diseases in their new hosts, in great part owing to the lack of adaptation that tends to moderate the toll that the virus imparts on the host. Examples of emergent human viruses thought to have come from animal reservoirs include *Human immunodeficiency virus* (HIV), thought to have emerged from a progenitor virus in chimpanzees (Gao et al. 1999); the novel coronavirus associated with severe acute respiratory syndrome (SARS), thought to have emerged from an animal reservoir (Guan et al. 2003); and the hemorrhagic viruses Ebola and Marburg (Leroy et al. 2005; Towner et al. 2006). Following the introduction into humans, emergence of all three of these viruses, especially HIV, also was facilitated by long-distance transport of humans (or monkeys in the case of Marburg virus). Human and other animal viruses also can be introduced via insect vectors, and the emergence of the West Nile virus in the USA is an example of an Old World virus that was introduced into the New World, with subsequent spread into bird, animal and human hosts mediated by indigenous mosquito vectors (Lanciotti et al. 1999).

In the case of plant viruses, the appearance of emergent viruses is usually mediated via an insect vector. However, with increasing global trade, the emergence of a virus in a new geographical region may be initiated by the introduction of infected plant materials (e.g., plants, propagative materials or seeds). Once introduced, the successful emergent virus expands into a new niche via activity of an existing insect vector or, less frequently, through spread by physical contact.

Finally, new forms of animal, human and plant viruses also emerge through common mechanisms of genetic variability, including mutation, reassortment and recombination. A classic example of this is the influenza virus, which can rapidly generate new strains with significantly altered virulence, via reassortment and recombination. Hence, the current worldwide apprehension that the "bird flu" strain of influenza (H5N1) will mutate and emerge as a virus with a greater capacity to be spread among the human population (Li et al. 2004).

3.2 What are Some Plant Viruses that Presently are Considered as Emergent?

Emergent plant viruses can be placed into two broad categories: entire groups of viruses (e.g., genera or families) or individual viruses. Some examples of plant virus groups that are emergent, on a global level, include (1) the whitefly-transmitted begomoviruses (genus *Begomovirus*, family *Geminiviridae*), (2) the thrips-transmitted tospoviruses (genus *Tospovirus*, family *Bunyaviridae*) and (3) the criniviruses (genus *Crinivirus*, family *Closteroviridae*). Some individual viruses that have emerged relatively recently include the potexvirus, *Pepino mosaic virus* (PepMV), an emergent tomato virus; and the sobemovirus, *Rice yellow mottle virus*. In addition, there also can be outbreaks of "new" viruses, such as those causing necrosis-associated diseases of tomato in Mexico, Spain and Guatemala (Verbeek et al. 2007). Another example is the emergence of a novel whitefly-transmitted potyvirus infecting cucurbits in Florida (Adkins et al. 2007).

3.3 What Factor(s) Lead to the Emergence of a Plant Virus?

There can be many factors that facilitate the emergence of a plant virus. These include genetic mechanisms such as random mutations, recombination, reassortment; long-distance movement to new agroecosystems; changes in vector population dynamics; and acquisition of novel viruslike entities. Quite often, the emergence of a plant virus involves more than one of these mechanisms (Table 3.1). In the rest of this chapter, we will consider these mechanisms and provide examples of viruses that have used these mechanisms in their emergence.

3.3.1 Long-Distance Movement

3.3.1.1 *Pepino mosaic virus*: Emergence of a Virus via Seed Dissemination

A new viral disease of tomato appeared in greenhouse-produced tomatoes in the Netherlands in 1999. The symptoms of this disease were variable, but included various degrees of mosaic (including a bright yellow mosaic) in leaves, distorted leaf growth and mottling in fruits. In some cases, plants would senesce prematurely and take on a grayish appearance (referred to as "thistletop"), whereas in other cases (conditions favorable for plant growth) plants would be symptomless. The causal agent was identified as PepMV, a member of the genus *Potexvirus*, plus-sense single-stranded RNA viruses with flexous rod-shaped virions. PepMV was not a new virus; it was first identified in Peru in 1974 infecting pepino (*Solanum muricatum*), a local solanaceous plant with a sweet-tasting fruit. Shortly after its identification in the Netherlands, the disease began to appear throughout Europe (e.g., France, Italy, Spain and the UK). In 2001, it was reported from the USA, where it has been increasing in incidence ever since. The disease is not particularly devastating to tomato production, causing losses of approximately 5–15%, much of which is due to reduced fruit quality. However, the virus is rapidly spread via mechanical means (i.e., touch, pollination, grafting and pruning) and it is extremely persistent owing to the stability of the virions. Thus, PepMV clearly fits the definition of an emergent virus.

One of the big questions was how did PepMV move from South America to Europe. Analysis of the nature of the initial outbreaks of the disease revealed an association with certain seed lots. Subsequent studies provided evidence of the association of the virus with seed, although at low levels and probably as an external contaminant. However, given how rapidly the virus can spread, plant-to-plant,

Table 3.1 Examples of emergent viruses/virus groups and mechanisms associated with their emergence

	Mechanism of emergence					
Virus/virus group	Long-distance movement	Changes in insect vector	Appearance of new species	Reassortment/ recombination	Acquisition of novel agents	Novel virus
Whitefly-transmitted begomoviruses	X	X	X	X	X	
Criniviruses	X	X	X			
Tospoviruses	X	X	X			
Pepino mosaic virus (potexvirus)	X			X		
Plum pox virus (potyvirus)	X			X		
Tomato torrado virus						X

it would not require high rates of seed contamination for the virus to become established in a tomato production system. Another factor that probably played a role in the emergence of PepMV is the practice of producing hybrid tomato seeds in countries such as Peru, which have favorable weather conditions and low labor costs. Thus, it is possible that PepMV was carried into Europe on hybrid tomato seed produced in Peru. If this were the case, then it would be expected that isolates of the virus from Peru, Europe and North America would be closely related. Indeed, studies of the population genetics of PepMV revealed high levels of nucleotide sequence similarity among tomato-infecting isolates from Peru, Europe and North America; consistent with emergence of a single genetic type (Verhoeven et al. 2003; Pagan et al. 2006). On the other hand, strains of the virus differing in genetic and biological properties (e.g., capacity to induce symptoms in tomato) also have been identified in Europe and elsewhere. There is evidence that these strains also may have originated from Peru, suggesting multiple introduction events into Europe. Finally, mixed infections of PepMV strains were detected in Spain, and recombinant isolates were detected (Pagan et al. 2006). Management of PepMV will require use of pathogen-free seed (facilitated by development of effective seed treatments or seed assays) and strict sanitation in tomato greenhouses; development of PepMV-resistant tomato varieties should be a long-term goal.

3.3.1.2 *Tomato yellow leaf curl virus*: Introduction of an Old World Virus into the New World

Tomato yellow leaf curl disease (TYLCD) was first described in the Middle East around 1940 (Cohen and Antignus 1994). In 1991, an isolate of the causal begomovirus, *Tomato yellow leaf curl virus* (TYLCV), from Israel was characterized and shown to possess a monopartite genome (Navot et al. 1991). Evidence that this single DNA component comprised the viral genome came from the development of TYLCD following introduction of the viral DNA into tomato plants via agroinoculation, a method where the plant pathogenic bacterium *Agrobacterium tumefaciens* is used to deliver the viral DNA instead of the whitefly vector (*Bemisia tabaci*). DNA sequence analysis revealed that the genome organization of TYLCV was similar to that of the DNA-A component of the bipartite begomoviruses, but with an extra open reading frame (ORF) on the virion-sense DNA strand (V1 ORF). With use of the cloned viral DNA as a probe and the sequence to generate TYLCV-specific PCR primers, TYLCV was found distributed throughout the Middle East (e.g., Israel, Egypt, Jordan, Lebanon and Cyprus). Now, over 15 years since the characterization of this TYLCV isolate (TYLCV-IL, now considered as the "type" isolate), this virus has spread to the New World, where it has emerged as a major constraint on tomato production. Here, we examine how this has happened.

TYLCV-IL was introduced into the Dominican Republic in the early 1990s (Salati et al. 2002). The identification of TYLCV in this Caribbean island country was the first report of the emergence of an Old World monopartite begomovirus in the New World. TYLCV, like all geminiviruses, is not seed-transmitted, and the whitefly vector is not capable of transcontinental flight. Thus, it was not surprising that anecdotal reports indicated that tomato transplants had been imported into the Dominican Republic from Israel in the early 1990s, owing to destruction of local transplants by heavy rains. Unfortunately, this inadvertent introduction was all TYLCV needed to gain a foothold in the Western Hemisphere. The Dominican Republic was well suited for the establishment of TYLCV: hot dry weather, overlapping crops of susceptible tomato varieties and large populations of the whitefly vector.

It did not take long for the virus to spread throughout the northern and southern processing tomato regions of the Dominican Republic, threatening to destroy an industry that had made the island self-sufficient for this commodity (Salati et al. 2002). Furthermore, the virus spread to Cuba, Jamaica, Puerto Rico and other islands in the Caribbean. In 1999, TYLCV was first reported in the USA from Florida, where it was discovered in tomato plants sold at retail stores. The inoculum source was hypothesized to be viruliferous whiteflies blown into Florida via high winds (Polston et al. 1999). Subsequently, TYLCV was reported from Georgia, Louisiana and even from North Carolina. More recently, TYLCV has emerged in northern Mexico (the states of Sinaloa and Tamulipas), where it caused major losses to fresh and processing tomato production in the state of Sinaloa during the 2005–2006 growing season (Brown and Idris 2006). The virus was subsequently identified in Texas and Arizona at the end of the 2006 growing season and, in March 2007 it was first identified in California (Rojas et al. 2007). TYLCV was also identified in Guatelmala in 2006.

DNA sequence analyses have confirmed that the TYLCV isolates throughout the New World are isolates of TYLCV-IL (i.e., total sequences more than 95% identical). Thus, since the initial introduction into the New World in the early 1990s, TYLCV has emerged, in a relatively short period of time, as a serious constraint on tomato production throughout the Caribbean, northern Mexico and the southern USA. While the factors mediating the rapid spread of this damaging pathogen in the Western Hemisphere are not completely understood, it is likely that it reflects a combination of migratory movements of the whitefly vector, together with movement of infected tomato transplants and/or other plants that are infected with TYLCV or harbor viruliferous whiteflies. The emergence of TYLCV represents an excellent example of the potential dangers associated with global movement of plant materials.

3.3.2 Emergence of Insect Vectors Precedes and Mediates Emergence of New Viruses from Pools of Viral Genetic Diversity in Reservoir Hosts

The *Begomovirus* and *Tospovirus* genera have emerged through a remarkable proliferation of new viral species over the past 10–20 years (Varma and Malathi 2003; Fauquet and Stanley 2003; Rojas et al. 2005; Whitfield et al. 2005; Prins and

Goldbach 1998). In both cases this was preceded by increases in the distribution and population of a polyphagous insect vector, often via the same mechanisms that facilitate long-distance dispersal of viruses. In the case of the begomoviruses, the vector is the sweet potato whitefly, *B. tabaci* (Brown et al. 1995); whereas for the tospoviruses, it is the western flower thrips (WFT), *Frankliniella occidentalis* (Prins and Goldbach 1998; Whitfield et al. 2005). Another factor that has contributed to the emergence of these viruses is the apparent availability of a pool of progenitor viral genetic information present in reservoir hosts (e.g., weed or native plants). Finally, the interaction of these vector–virus combinations is mediated by increasing land conversion and intensification of agricultural production, e.g., through the use of new high-yielding varieties, improvements in irrigation and irrigation technology (especially drip irrigation), and increased use of pesticides and fertilizers (Matson et al. 1997). Indeed, it is often at the interface of land conversion/agricultural intensification where emergence of new viruses commonly occurs.

3.3.2.1 Emergence of Thrips and Tospoviruses

One of the definitions of an emerging group of viruses is the relatively rapid emergence of new virus species in diverse geographic locations. RNA viruses of the genus *Tospovirus* are the only plant-infecting viruses in the family *Bunyaviridae*. The type member *Tomato spotted wilt virus* (TSWV) has been known since 1919, but it is only recently that new tospovirus species have proliferated (Whitfield et al. 2005). Thus, the tospoviruses qualify as a group of emergent viruses on the basis of the fact that new species are being described from a range of host plants in a diversity of geographical locations (Prins and Goldbach 1998; Whitfield et al. 2005). This proliferation has been correlated with tremendous increases in thrips populations, especially the polyphagous WFT, which has been mediated by agricultural intensification (Prins and Goldbach 1998). In the case of thrips, overuse or reliance on a few insecticides has also led to the generation of insecticide-resistant populations. Together with the difficultly of applying insecticides to flowers and other places where thrips feed and reproduce, this has limited the effectiveness of vector control for management of these tospovirus diseases. Most of the new tospovirus species have originated in tropical regions of Asia, perhaps suggesting a "hot spot" of viral genetic diversity in reservoir hosts. However, two species are emerging as significant economic threats in temperate North America: *Impatiens necrotic spot virus* (INSV) and *Iris yellow spot virus* (IYSV).

INSV was first discovered as a virus affecting ornamental plants in greenhouses. It has become widely distributed in the USA, presumably via the movement of infected ornamental plants and/or in viruliferous thrips carried on ornamentals (Daughtrey et al. 1997). Thus, INSV has emerged as an important virus in the ornamental plant industry. More recently, INSV has been reported causing diseases in a number of crop plants. For example, INSV was identified as the cause of tomato spotted wilt-like symptoms in peppers in Georgia and lettuce in California (Naidu et al. 2005; Koike, Kuo and Gilbertson, unpublished data). It remains to be seen whether INSV will become an emergent virus of vegetable crops. The availability of rapid detection methods such as immunostrips for TSWV and INSV will greatly facilitate identification and monitoring of these tospoviruses.

Another emergent tospovirus is IYSV, which infects onions and other members of the onion family (Gent et al. 2006). It fits the description of an emergent virus because it initially had a limited distribution but, more recently, it has been reported from a wide range of geographic locations. Although the nature of the emergence of IYSV is not clear, the recent plethora of reports of the virus may relate to improved diagnosis, including better recognition of the symptoms (i.e., the chlorotic diamond-shaped lesions on leaves and stems) and availability of serological detection tools (Gent et al. 2006). The potential for IYSV to cause losses to onion seed production, through the lodging of seed stalks at points with virus lesions, makes this an important emerging virus in the onion industry.

Finally, the finding of limited genetic diversity among isolates of INSV and IYSV from different hosts and geographic locations seems to support an initial emergence event followed by effective long-distance dispersal. This dispersal has undoubtedly been favored by the fact that the key hosts of INSV and IYSV are vegetatively propagated, thereby avoiding the limitation imposed by the lack of efficient tospovirus seed transmission.

3.3.2.2 *Bemisia tabaci*: an Insect Vector that has Mediated the Emergence of Begomoviruses and Criniviruses

There is perhaps no greater culprit in the emergence of new plant viruses than the sweet potato whitefly, *B. tabaci* biotype B (=silverleaf whitefly, *Bemisia argentifolii*). Thought to have originated in the Middle East/Asia, this insect has emerged as a major pest of vegetable production worldwide (Brown et al. 1995). Though a significant pest in its own right, it has mediated the emergence of the begomoviruses and the criniviruses over the past 20 years.

Over the past 20 years, *B. tabaci* (biotype B as well as other biotypes) has mediated the emergence of well over 100 new begomovirus species. This has resulted in the whitefly-transmitted begomoviruses supplanting potyviruses as the group of plant viruses with the largest number of recognized viral species. In this chapter, we will provide examples of the mechanisms involved in the emergence of new begomovirus species (Fig. 3.1). For detailed reviews of this subject, the reader is referred to other reviews (Polston and Anderson 1997; Seal et al. 2006; Varma and Malathi 2003).

The worldwide emergence of begomovirus diseases has been facilitated, in part, by parallel evolution. Here, genetically distinct begomovirus species evolve in distinct geographical regions (e.g., continents) to cause the same or similar disease symptoms in a given crop plant (Fig. 3.1). This occurs as whiteflies introduce genetically diverse progenitor begomoviruses, existing in reservoir hosts in these geographically

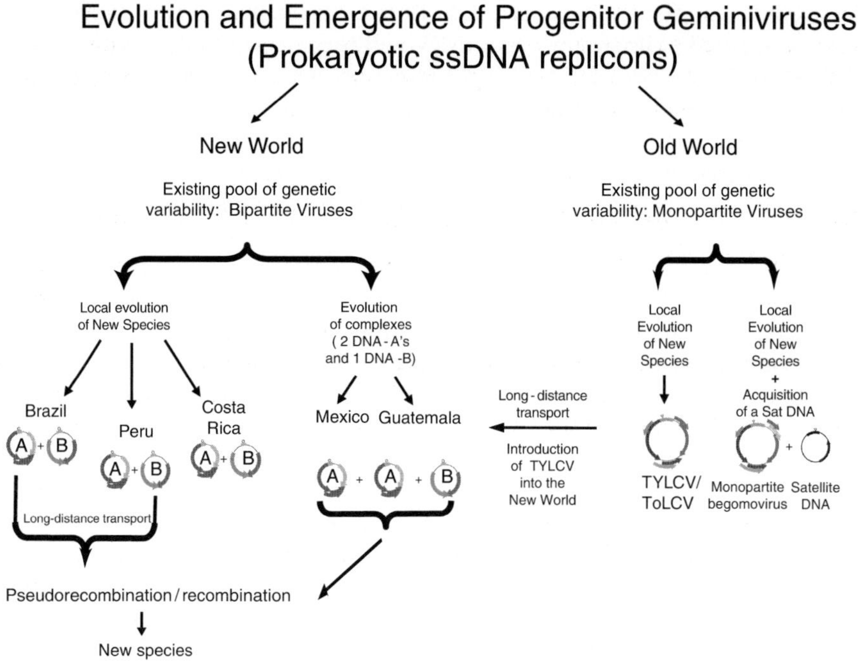

Fig. 3.1 Mechanisms of emergence of whitefly-transmitted begomoviruses (genus *Begomovirus*, family *Geminiviridae*). *TYLCV*, *Tomato yellow leaf curl virus*; *ToLCV*, *Tomato leaf curl virus*

distinct regions, into widely grown susceptible crops (e.g., common bean, cotton, cucurbits, peppers and tomato). This results in the simultaneous or parallel evolution of new crop-infecting begomovirus species in multiple locations. However, while the viruses are genetically diverse, the common mechanisms of begomovirus pathogenicity in these genetically uniform hosts results in the induction of similar disease symptoms.

A good example of this phenomenon is the disease bean golden mosaic. It has been established that two genetically distinct lineages of begomoviruses emerged in South America (Brazil) and Mesoamerica, each of which causes the disease bean golden mosaic (Gilbertson et al. 1993a). Evidence that viruses representing these two lineages are genetically distinct includes sequence divergence (approximately 80% identity), inability to generate infectious reassortants, and differences in sap transmissibility and germ plasm susceptibility. On the basis of these differences, viruses representing these lineages were named *Bean golden mosaic virus* (the initially described virus from Brazil) and *Bean golden yellow mosaic virus* (a cluster of related Mesoamerican viruses). This diversity can complicate breeding for disease resistance and deployment of resistant varieties (i.e., when a variety has resistance against only a subset of the viruses that cause the same or similar symptoms). A similar situation has since been described for begomoviruses associated with other diseases, including African cassava mosaic, cotton leaf curl and tomato leaf curl (Varma and Malathi 2003). This situation has undoubtedly been facilitated by the emergence of the polyphagous *B. tabaci* biotype B, which feeds on a wide range of plants and, thus, would have a higher probability to acquire, mix and deliver a diversity of begomovirus components into potential new hosts.

3.4 Reassortment and Recombination: Effective Mechanisms of Variability for DNA Viruses

In terms of viral evolution, it is widely held that geminiviruses are at a disadvantage when compared with their RNA virus counterparts because of the lower frequency of mutations introduced during replication of the viral genome. However, it is now clear that geminiviruses have made up for this deficiency, if it exists, through reassortment and recombination (Padadim et al. 1999; Rojas et al. 2005). In the geminiviruses, recombination is facilitated by a type of rolling circle replication referred to as recombination-dependent replication, which favors recombination and generates a diversity of viral DNA forms (Preiss and Jeske 2003).

Initial evidence for recombination in bipartite begomoviruses came from studies in which deleterious mutations were repaired by replacement with wild-type DNA sequences from the same species (Roberts and Stanley 1994). It was further hypothesized that the bipartite genome might facilitate recombination between components, even among distinct viral species. However, initial studies showed a limited capacity for reassortment among components of begomoviral species, owing to the specificity of the replication-associated (Rep) protein for the cognate origin of replication [common region (CR) of the components of a bipartite begomovirus]. The formation of infectious reassortants between the cloned DNA components of closely related bipartite begomovirus species *Bean dwarf mosaic virus* (BDMV) and *Tomato mottle virus* (ToMoV) suggested that specificity of the Rep protein was less stringent than initially thought (Gilbertson et al. 1993b). Additional reports confirmed these findings, and revealed that reassortment was common among closely related begomoviruses; this even was reported to occur between distantly related bipartite begomoviruses (i.e., from different phylogenetic clades; Garrido-Ramirez et al. 2000).

Key insight into how these processes could mediate the emergence of new bipartite begomoviruses came from results showing that passage, through plants, of a less fit reassortant (ToMoV DNA-A/BDMV DNA-B) led to emergence of a more pathogenic virus through a recombination event that resulted in the exchange of the ToMoV DNA-A CR with that of the BDMV DNA-B component. This allowed the emergence of a fitter bipartite begomovirus, in large part due to the increased replication of the BDMV DNA-B (Hou and Gilbertson 1996). The question remained about the relative importance of these mechanisms in nature. Compelling evidence for a role of CR exchange in begomovirus evolution in nature came from a study of cassava-infecting begomoviruses in India. Here, a new bipartite begomovirus, *Sri Lankan cassava mosaic virus* (SLCMV), was identified that appeared to have emerged from a monopartite begomovirus (SLCMV) that acquired a DNA-B component from another bipartite begomovirus (*Indian cassava mosaic virus*) by CR exchange (Saunders et al. 2002).

3.4.1 Recombination

TYLCV is clearly a virus that is well adapted to its primary host, tomato. While it is able to infect other plants, including certain crop plants (e.g., peppers and common bean), ornamentals (e.g., lisianthus) and weeds, these infections are often symptomless and associated with low viral titers (Salati et al. 2002). The virus either lacks a factor needed for effective colonization of these non-tomato hosts, or the hosts possess some defense factor(s) or response(s) that prevents efficient infection. If selective pressures are placed on TYLCV that influence the capacity of the virus to infect tomato (e.g., the introduction of a resistant variety or an extensive tomato-free period), the virus may respond by genetic variation, leading to the emergence of a new variant with a broader host range.

Indeed, sequence analyses of TYLCV isolates from around the world have revealed evidence of extensive recombination (Fauquet and Stanley 2003). The first isolate of TYLCV to be fully characterized was the previously mentioned TYLCV-IL (Navot et al. 1991). A second isolate from Israel, TYLCV-Mld[IL], was subsequently described; it was a recombinant, having a genome comprising approximately 75% TYLCV-IL and approximately 25% of an unknown begomovirus. The recombinant portion of the genome included the intergenic region and the 5′ end of the C1 ORF, which encodes the Rep protein. Subsequently, epidemics of TYLCD in southern Italy in the late 1980s were shown to be caused by a distinct begomovirus species (less than 80% identical to TYLCV), and it was named *Tomato yellow leaf curl Sardinia virus* (TYLCSV). The emergence of TYLCSV likely reflected an independent parallel evolution event mediated by increases in whitefly populations in southern Italy.

In 1992, TYLCD appeared in southern Spain and, consistent with geographic proximity to Italy, the causal agent was a strain of TYLCSV (TYLCSV-ES). In 1997, TYLCV-Mld[IL] was detected in southern Spain and, in 1999, a recombinant TYLCV was identified that was composed equally of TYLCSV-ES and TYLCV-Mld[IL] sequences. This recombinant, named TYLCMalV, was hypothesized to have emerged because of a selective advantage over the parental viruses in terms of a wider host range, more efficient whitefly transmission and being more infectious in TYLCV-resistant tomato varieties (Monci et al. 2002). Evidence for the selective advantage of this recombinant came from results of a study of the population structure of begomoviruses associated with TYLCD in southern Spain from 1999 to 2003 indicating that TYLCMalV had become established and had spread throughout this important tomato production region (Garcia-Andres et al. 2007). In addition, mixed infections of TYLCV were common, and a new type of recombinant, between the type strain of TYLCV and TYLCSV-ES, was detected. This clearly demonstrates how recombination can generate novel begomovirus forms that can allow for adaptation and expansion into new environments, and that can compete with existing viruses.

3.4.2 *Reemergence of Cassava Mosaic Disease in Africa: a Role for Reassortment and Recombination*

Further evidence that these mechanisms play a key role in emergence of new begomoviruses has come from analysis of begomoviruses associated with the reemergence of cassava mosaic disease (CMD) in Africa. This disease has long been known on the African continent, and causes significant economic losses to production of this staple crop. Multiple begomovirus species can cause CMD, and these likely arose via parallel evolution. On the basis of the biology of these viruses and the nature of cassava cultivation, management strategies were implemented, including use of virus-free propagative material, deployment of resistant varieties and roguing of infected plants. This approach seemed to keep the disease at manageable levels (Legg 1999).

However, in the 1980s the incidence and severity of CMD increased markedly in East Africa (Legg 1999; Legg and Fauquet 2004). This was associated with the emergence of highly virulent forms via reassortment and recombination. For example, recombination between *East African cassava mosaic virus* (EACMV) and *African cassava mosaic virus* (ACMV), in which capsid protein (CP) gene sequences of ACMV were exchanged with homologous sequences in EACMV, has given rise to a highly virulent recombinant (EACMV-UG2) that has been implicated in these disease outbreaks. In addition, reassortment between other recombinant EACMV components has led to the emergence of other highly virulent forms in other parts of southeast Africa (Pita et al. 2001). Together with increases in whitefly populations on cassava, these emerging viruses pose a major threat to cassava production on the African continent.

3.5 Tripartite Begomovirus Complexes: A Way for Bipartite Begomoviruses To Fight Host Defense Responses?

The bipartite begomovirus genome has evolved to allow for efficient replication and spread of a single-stranded DNA replicon in plants, in addition to providing opportunities for viral variability (Rojas et al. 2005). Thus, attempts to generate viable viruses having dramatically altered genome size, or with more than two components, have generally been unsuccessful. However, there is increasing evidence that complexes of more than two bipartita DNA components have emerged as primary causal agents of diseases such as chino del tomate in Mexico, leaf curl of tomato in Guatemala and African cassava mosaic in Uganda (Fig. 3.1). Because such complexes are typically not viable, there must be a selection pressure to generate and maintain these specific complexes.

Evidence for the existence of this type of synergistic interaction among bipartite begomovirus components came from the observation that the tripartite combination of *Pepper huasteco yellow vein virus* (PHYVV) DNA-A and DNA-B and *Pepper golden mosaic virus* (PepGMV) DNA-A induced strikingly more severe disease symptoms in *Nicotiana benthamiana*, tomato and pepper plants compared with symptoms induced by PHYVV alone (Sharp et al. 1999). The specificity of this interaction was demonstrated by the consistent detection of all three components in plants with the severe symptom phenotype, and the inability to form such a complex with the DNA-A component of another bipartite tomato-infecting begomovirus from Mexico. Furthermore, this complex was detected in the field in Mexico, consistent with its emergence and transmission by *B. tabaci* (Mendez-Lozano et al. 2003). While the mechanism underlying the synergism among these begomovirus components is not known, the CP was not involved because the synergism was not altered with a PepGMV DNA-A CP mutant.

Possible insight into the mechanism underlying these synergistic interactions comes from analyses of complexes associated with CMD in Uganda (Pita et al. 2001). Here, evidence has been provided that the product of the AC4 ORF can function as a suppressor of gene silencing, and that the relative effectiveness of the suppression varies among DNA-A components. Thus, the selection pressure to maintain more than one DNA-A component may well reflect the need to have a compliment of AC4 suppressors, thereby allowing for effective suppression of this powerful antiviral defense response (Vanitharani et al. 2005).

3.6 Acquisition of Novel Viruslike Entities: Monopartite Begomoviruses and their Satellite DNAs

Old World monopartite begomoviruses can be associated with a DNA satellite or "extrachromosomal" DNA. In studies of the etiology of a number of Old World geminivirus diseases [e.g., Ageratum yellow vein in *Ageratum conyzoides* (Singapore), Bhendi yellow vein mosaic in okra (India), cotton leaf curl in cotton (Pakistan), and tomato and tobacco leaf curl in tomato and tobacco, respectively (China)], a single begomovirus DNA component was detected, suggesting the disease was caused by a monopartite begomovirus (as had been shown for TYLCV). However, when introduced back into the natural hosts, the cloned DNA component was infectious but induced symptoms that were much milder than those observed in plants in the field (Mansoor et al. 2003). This puzzling observation was explained following the identification of sub-virus-sized satellite DNAs (satDNAs) associated with these diseases. When these satDNAs were introduced into plants, together with their cognate monopartite begomovirus, the characteristic disease symptoms were induced. These satDNAs, also referred to as DNA-βs, are approximately 1.4 kb (about half the size of the begomovirus genome) and require the helper begomovirus for replication and movement. The satDNAs share no sequence similarity with the helper virus except a stem-loop structure (presumably the origin of replication), and have a single ORF (βC1). The βC1 protein is a symptom determinant, as an intact *βC1* gene is required for satDNA to mediate symptom development, and expression of βC1 in transgenic plants results in a symptomatic phenotype (Saunders et al. 2004; Saeed et al. 2005).

What is emerging from studies of Old World begomovirus-associated diseases is that most of these diseases are complexes of a monopartite begomovirus and a satDNA (Fig. 3.1). This conclusion is further supported by the identification of tremendous diversity of DNA-β satDNAs (more than 130 sequences in the GenBank) from a range of geographic locations (e.g., Africa, China, India, Indonesia, and Pakistan; Briddon et al. 2003). This level of diversity is consistent with the idea that this "unholy alliance" was a relatively ancient event, which has facilitated the emergence of monopartite begomoviruses as important pathogens of a range of crop plants in the Old World (Rojas et al. 2005). It is not clear where the satDNA originated from, but it has been suggested that it came from another, yet to be characterized, type of single-stranded extrachromosomal DNA (Mansoor et al. 2003).

The acquisition of the satDNA may be analogous to the acquisition of the DNA-B component, which facilitated the emergence of the bipartite begomoviruses in the New World. However, while it is well established that the bipartite begomovirus DNA-B component encodes two proteins required for movement (Rojas et al. 2005), the function of the satDNA is less clear. One key function may be the suppression of host defenses, such as gene silencing. Consistent with this hypothesis, functional analyses of the *βC1* gene of some satDNAs have revealed that it is a suppressor of gene silencing (Cui et al. 2005; Kon et al. 2007). This may be an important function, as gene silencing has been shown to target begomoviruses (Bisaro 2006; Rojas et al. 2005; Vanitharani et al. 2005). However, it is also possible that the satDNA and the βC1 may be involved in other functions, such as movement.

An important characteristic of these satDNAs that differentiates them from the bipartite begomovirus DNA-B component is their capacity to be maintained (replicated, moved and encapsidated) by multiple monopartite begomoviruses (Mansoor et al. 2003). This promiscuity has revealed more flexibility in Rep-protein-mediated replication than had been previously thought. Moreover, the capacity to be replicated by multiple begomoviruses means that satDNAs can form complexes with multiple monopartite begomoviruses, thereby allowing these viruses to increase their incidence and/or host range. Thus, it is clear that there is a strong selective advantage for monopartite begomoviruses to "partner" with a satDNA, and that this has had a major impact on the emergence of the begomoviruses in the Old World.

Symptoms induced by RNA viruses also can be moderated by satellite RNAs (satRNAs; Simon et al. 2004), such as the satRNAs associated with *Cucumber mosaic virus* (CMV). Like the begomovirus satDNA (e.g., DNA-β), the CMV satRNAs require the helper virus (CMV) for replication, and have little sequence similarity with the helper virus. In contrast to the begomovirus satDNA, the CMV satRNA are linear single-stranded RNAs, apparently do not encode for any proteins, and usually attenuate symptoms induced by the helper virus (Simon et al. 2004). However, certain CMV satRNAs increase disease symptoms, inducing either necrosis or chlorosis. The induction of necrosis has been shown to be associated with the initiation of programmed cell death mediated by the satRNA minus-strand (Simon et al. 2004). The origins of begomovirus satDNAs and CMV satRNAs remain a mystery; however, whereas the begomovirus satDNAs are commonly found in nature, the CMV satRNAs tend to be found in experimental systems. Thus, the begomovirus/satDNA combination is a synergistic interaction, whereas the CMV/satRNA interaction seems to reflect a situation where the satRNA is a parasite of the virus.

3.7 Emergence of Diseases Caused by Novel Viruslike Agents

Though many emerging plant virus diseases are associated with previously characterized viruses or viruslike agents, some are caused by novel agents. For example, necrosis-associated diseases of tomato in Mexico [*marchitez manchada* (Sinoloa spotted wilt)], Spain [*torrado* (burned or roasted) disease; Verbeek et al. 2007] and Guatemala [*mancha de chocolate* (chocolate spot)] have emerged and appear to be caused by a novel virus or viruses. The symptoms of these diseases appear similar to those induced by the tospovirus TSWV; however, tests for known tomato-infecting viruses have given negative results. The diseases are also graft- and sap-transmissible, consistent with a viral etiology. It was reported recently that the torrado disease is caused by a novel plant picorna-like virus, most closely related to single-stranded RNA viruses in the genera *Sequivirus*, *Sadwavirus* and *Cheravirus*. The name proposed for this virus is Tomato torrado virus (Verbeek et al. 2007). It is not clear whether the necrosis-associated diseases in Guatemala and Mexico are caused by Tomato torrado virus or some other virus. It is also not clear what the vector(s) of these agents is. If these diseases turn out to be caused by the same or closely related viruses, their recent emergence in New and Old World locations may suggest a long-distance transport mechanism, e.g., in association with seeds or propagative materials.

3.8 Multiple Mechanisms often Underlie the Emergence of Plant Viruses

3.8.1 Long-Distance Movement and an Emergent Vector: Cucurbit yellow stunting disorder virus

The criniviruses represent a group of viruses that have emerged over the past 10–20 years in association with the worldwide emergence of whiteflies (Wisler et al. 1998). Another factor that has contributed to the emergence of these viruses is the recognition of their association with yellowing-type symptoms that were previously attributed to nutrient deficiencies. *Cucurbit yellow stunting disorder virus* (CYSDV) is a crinivirus that induces a striking interveinal yellowing of various cucurbits (cucumbers and melons) and is transmitted by *B. tabaci* (Celix et al. 1996; Wisler et al. 1998). The virus emerged as a serious constraint on cucurbit production in the Middle East and Mediterranean regions in the early 1990s, and this was associated with the displacement of the greenhouse whitefly (*Trialeurodes vaporariorum*) by *B. tabaci*. In 2000, CYSDV was reported in Texas and subsequently in Guatemala. In 2006, a significant outbreak of CYSDV occurred in Arizona, California and northern Mexico (Kuo et al. 2007). Thus, CYSDV represents another example of an Old World virus being introduced into the New World. Moreover, as criniviruses are not seed-transmitted and CYSDV infects annual host plants, the virus was likely introduced via infected plants or viruliferous whiteflies carried on plants (hosts or nonhosts of CYSDV). Analyses of genetic diversity among CYSDV strains from different regions, as well as from a single region over an 8-year period, revealed a relatively homogenous population (Marco and Aranda 2005). Thus, CYSDV probably has emerged via long-distance transport followed by local spread mediated by existing *B. tabaci* populations.

3.8.2 Mutation, Recombination and Long-Distance Movement: Plum pox virus

Sharka disease, caused by the potyvirus *Plum pox virus* (PPV), is one of the most damaging diseases of *Prunus* spp. (e.g., peach, apricot, nectarine, plum and sweet and sour cherry). Prior to the 1990s, it was a disease exclusively found in the Old World (e.g., Europe and Asia); however, in 1994 it was detected in Chile and in 2000 it was detected in Canada and the USA (Candresse and Cambra 2006). Despite extensive quarantine efforts, this long-distance movement likely occurred through movement of infected propagative materials, with subsequent spread via aphids. Genetic analyses of PPV isolates, from a diversity of hosts and locations, have revealed extensive genetic variability, with at least six subgroups (strains or serotypes) recognized. Some of this diversity can undoubtedly be attributed to random mutation, mediated by the lack of proofreading activity of the viral replicase; however, at least one of these subgroups, PPV-Rec, emerged via recombination events between isolates of other subgroups (e.g., PPV-D and PPV-M; James and Glasa 2006). Extensive efforts are under way to eradicate this emergent virus from Canada and the USA as well as to develop resistant varieties of *Prunus* spp.

3.9 Bringing Them All Together: Tomato Yellow Leaf Curl/Leaf Curl Disease in West Africa

Tomato yellow leaf curl disease (TYLCD) and tomato leaf curl disease (ToLCD) have emerged as major constraints on tomato production in West Africa, including countries such as Mali, Benin, Burkina Faso and Senegal. The symptoms in infected plants include stunted and distorted growth, and varying degrees of leaf curl and crumple, chlorosis and purpling. Recent investigation of these diseases has revealed a complex etiology, involving emergence of multiple new begomovirus species, a recombinant virus and a satDNA. Thus, three new monopartite begomovirus species have been associated with this epidemic: *Tomato leaf curl Mali virus* (ToLCMLV), *Tomato yellow leaf curl Mali virus* (TYLCMLV) and *Tomato yellow leaf crumple virus* (ToYLCrV). Analysis of the complete sequence of an infectious TYLCMLV clone revealed it was a recombinant virus, with the genome comprising approximately 80% TYLCV-IL, with approximately 20% from an uncharacterized begomovirus. The recombinant region of the genome included the intergenic region (up to the nicking site in the origin of replication) and the 5′ end of the C1 ORF, which encodes the Rep protein. In addition, a novel approximately 1.4 kb satDNA was cloned from TYLCMLV-infected plants and, when coinoculated with TYLCMLV, it increased symptom severity in tomato, common bean and *N. glutinosa*. With use of PCR and virus-specific primers, it was established that tomatoes with severe stunting and distortion symptoms in West Africa were associated with mixed begomovirus infection. Thus, the TYLCD/ToLCD epidemic in West Africa is caused by a complex of locally emergent begomoviruses, a satDNA and perhaps a progenitor virus (TYLCV) that was previously introduced into West Africa. Management of this disease complex will be challenging and will likely require an integrated approach (Rojas et al. 2005).

3.10 Emergence of a New Virus is not always Catastrophic: Failure of New Viral Diseases to Emerge Following a Major Change in the Whitefly Vector Population

3.10.1 A History of Whitefly-Transmitted Viruses in the Imperial Valley of California

While the emergence of whitefly-transmitted begomoviruses has led to the appearance of diseases that have caused significant economic losses to a wide range of crop plants throughout the world, this is not always the case. The Imperial Valley of California is an irrigated desert agricultural production area in southern California where a variety of vegetable and forage crops are grown, including alfalfa, carrots, cotton, melons and onions. This area historically has sustained significant populations of sweet potato whiteflies (*B. tabaci* biotype A). Two species of whitefly-transmitted bipartite begomoviruses, *Squash leaf curl virus* (SLCV) and *Cotton leaf crumple virus* (CLCrV) have also been described from this area; however, these viruses (or the diseases they cause) have been known for decades (since the early 1980s for SLCV and since the 1940s for cotton leaf crumple disease) and have not been a major economic problem. Thus, these viruses would not be classified as emergent.

However, in the early 1980s, a new whitefly-transmitted virus emerged that caused a severe yellows disease of lettuce and cucurbits. This virus was identified as closterovirus-like and, subsequently, it was shown to be a novel bipartite closterovirus. The virus was named *Lettuce infectious yellows virus* (LIYV), and it is the type member of the genus *Crinivirus* in the family *Closteroviridae* (Wisler et al. 1998). This emergent virus caused significant economic losses to lettuce and melon production in the Imperial Valley, and it was feared that LIYV could be a limiting factor for production of these crops.

In the early 1990s, this situation in the Imperial Valley changed with the introduction of the B biotype of *B. tabaci* (silverleaf whitefly). Massive outbreaks of whiteflies followed the introduction of this exotic insect (Perring et al. 1991; Toscano et al. 1998), and the B biotype quickly displaced the indigenous A biotype. These outbreaks resulted in crop losses owing to the sheer magnitude of the whitefly populations and their feeding-induced physiological abnormalities (e.g., irregular ripening of tomato and silverleaf of squash). Another big concern was that the insects would increase the incidence/severity of existing viruses and/or facilitate the emergence of new virus diseases. Unexpectedly, the incidence of LIYV decreased dramatically, and the infectious yellows disease essentially disappeared from the Imperial Valley. Whitefly transmission experiments, conducted with *B. tabaci* biotypes A and B provided the explanation: *B. tabaci* biotype B was an inefficient vector of LIYV. Thus, here is a rather unusual situation where introduction of an exotic insect has led to the reduction of an economically important emergent virus. Indeed, it appears that LIYV may well have become extinct in the Imperial Valley!

The B biotype is a known vector of begomoviruses, although it is generally a less efficient vector than the A biotype. However, the apparent reduced vectoring efficiency is compensated for by the biotype B feeding on a wider range of host plants and having a higher rate of reproduction. Thus, it was feared that the high populations of biotype B whiteflies would lead to the emergence of new begomoviruses. This could occur by introducing variants of indigenous weed-infecting begomoviruses into crop plants or mixtures of begomoviral components, resulting in the evolution of a new emergent crop-infecting virus via reassortment and recombination. This latter scenario has been proposed to explain the evolution of CLCrV (Idris and Brown 2004; Seo et al. 2006). However, this scenario has yet to unfold in the Imperial Valley, almost 20 years since the introduction of biotype B. Thus, the incidence of SLCV and CLCrV did not change significantly, nor did new crop-infecting begomoviruses emerge to threaten agricultural production. The fact that SLCV and CLCrV still appear, to some extent, every year, taken together with the subsequent introduction of CYSDV and TYLCV, indicates that the failure to observe the emergence of new begomoviruses was not due to an inability of the B biotype whiteflies to transmit viruses. More likely this reflects a limited genetic pool of indigenous viruses in this desert region (as compared with a tropical agroecosystem), or perhaps a lack of extensive cultivation of highly susceptible begomovirus hosts (e.g., tomato, pepper and common bean). Thus, the Imperial Valley agroecosystem appeared to lack a component necessary for the rapid and widespread emergence of new begomoviruses.

3.10.2 Emergence of CuLCrV in the Imperial Valley

In the fall of 1998, when watermelon volunteers at the edge of the agricultural production area of the Imperial Valley showed symptoms of geminivirus infection, it was thought to be an outbreak of SLCV. However, while tests confirmed begomovirus infection in these plants, sequence analysis of the PCR-amplified fragments of the virus genome revealed only approximately 85% sequence identity with SLCV, the most closely related previously characterized begomovirus. Subsequent cloning and sequencing of the DNA-A and DNA-B components of this begomovirus established that it was a new species, and it was named *Cucurbit leaf crumple virus* (CuLCrV; Guzman et al. 2000). Phylogenetic analyses confirmed that CuLCrV was a distinct begomovirus species in the SLCV cluster of New World bipartite begomoviruses. The close relationship of CuLCrV and SLCV was further demonstrated by the finding that infectious reassortants could be generated between the cloned DNA components of these viruses (Brown et al. 2002). Thus, while the source of CuLCrV remains unknown, it is thought to have evolved from a progenitor begomovirus infecting an indigenous host plant that is distributed outside the cultivated lands of the Imperial Valley. A similar progenitor virus may have given rise to SLCV over 20 years earlier. This hypothesis could be tested by surveying weeds and other indigenous hosts (symptomatic and symptomless) for the presence of begomovirus infection by PCR with degenerate begomovirus primers (Rojas et al. 2005).

In the years following its identification, CuLCrV emerged as the cause of leaf crumpling, curling and chlorosis symptoms in cantaloupe, watermelon and squash in the Imperial and Coachella Valleys of California, southern Arizona and northern Mexico. Therefore, CuLCrV represents the first emergent begomovirus that has appeared in the desert southwest following the outbreaks of whitefly biotype B. In the case of squash, the virus caused in the Coachella valley economic losses. However, observation of CuLCr disease development in cantaloupe and watermelon revealed that, while severe symptoms initially developed in infected plants, symptoms eventually became markedly attenuated as the plants continued to grow. Eventually, plants showed few or no symptoms and fields with affected plants provided acceptable yields. This "recovery from infection" occurred in cantaloupe and watermelon, but not in squash or pumpkin. It also was consistently reproduced in the laboratory. Analysis of recovered tissue revealed a considerable reduction of viral titer compared with nonrecovered symptomatic tissues, high rates of methylation of viral DNA and the presence of small CuLCrV-derived RNAs. Furthermore, reinoculation of recovered tissue with CuLCrV failed to result in symptom development, whereas inoculation of recovered tissues with CMV led to the return of severe symptoms and increased CuLCrV titers. This latter result suggests that the

2b silencing suppressor of CMV may have interfered with the host defense mechanism that attenuated symptoms of CuLCrV. Taken together, these results indicate that cantaloupe and watermelon stage a vigorous defense response against CuLCrV infection involving virus-induced gene silencing. Thus, though CuLCrV is a new emergent virus, the major economic hosts of the virus in the Imperial Valley mount a vigorous defense response, limiting the economic importance of the virus.

3.11 Conclusions

Over the past 10–20 years, groups of plant viruses as well as individual viruses have emerged as significant constraints on crop production worldwide. This emergence can involve a range of mechanisms (Table 3.1), depending on the virus(es) involved and the environment. Existing viruses can be moved long distances via human activities, allowing for subsequent establishment and spread in compatible agroecosystems. Irrespective of whether the virus has a DNA or an RNA genome, mechanisms of variability such as mutation, reassortment and recombination allow for the generation of new forms of existing viruses that have the potential to emerge as important pathogens. This is facilitated by new selection pressures placed on the viral population, such as those associated with modifications of existing agroecosystems. However, the rate at which such variants are generated and their economic impact is a function of multiple factors (efficiency of vector transmission, the existing pool of viral genetic variability in a region, nature of the host, aspects of the agroecosystem, etc.). Novel viruses or viruslike entities can also emerge, and these often appear at interfaces of agricultural and undeveloped lands. In the case of plant viruses, emergence of variants of known viruses or novel agents is greatly facilitated by increases or emergence of insect vector populations. Entire groups of viruses (e.g., the begomoviruses, criniviruses and tospoviruses) have emerged following the worldwide emergence of their insect vectors. Moreover, given the continued global movement of plant materials and seeds, expansion of agriculture into new areas, and the propensity of large-scale commercial agriculture to favor development of large populations of insect pests, it is highly likely that new plant viruses will continue to emerge.

The challenge in dealing with emergence of new viruses is significant, but it has been lessened by improvements in technology and increased understanding of viral genetic diversity, ecology and genetics. Thus, with new detection technologies (PCR and sequencing and microarray-based technologies), potentially new emergent viruses can be identified sooner. However, identification of novel viruses can still be very challenging, and often requires a combination of new technology and innovative approaches. Once an emerging plant virus has been identified, effective detection tools need to be developed and applied to answer questions about the biology of the virus. This information can then be used to develop effective management strategies. Ideally, multiple approaches will lead to development of an integrated pest management (IPM) approach. Such an approach has led to the effective management of the emergent TYLCV in the Dominican Republic, and tomato production has actually increased since the introduction of the virus (Salati et al. 2002). Similar management approaches are now being brought to bear on begomovirus diseases of tomato in West Africa and Central America. Thus, it is hoped that the use of IPM strategies, tailor-made for emergent viruses based on understanding of the biology of the virus, will minimize the impact of these viruses on world food production.

References

Adkins S, Webb SE, Achor D, Roberts PD, Baker CA (2007) Identification and characterization of a novel whitefly-transmitted member of the family *Potyviridae* isolated from cucurbits in Florida. Phytopathology 97:145–154

Bisaro DM (2006) Silencing suppression by geminivirus proteins. Virology 344:158–168

Briddon R, Bull S, Amin I, Idris A, Mansoor S, Bedford I, Dhawan P, Rishi N, Siwatch S, Abdel-Salam A, Brown J, Zafar Y, Malik KA, Markham PG (2003) Diversity of DNAβ, a satellite molecule associated with some monopartite begomoviruses. Virology 285:234–243

Brown JK, Idris AM (2006) Introduction of the exotic monopartite *Tomato yellow leaf curl virus* into West Coast Mexico. Plant Dis 90:1360

Brown JK, Frohlich DR, Rosell RC (1995) The sweetpotato or silverleaf whiteflies: biotypes of *Bemisia tabaci* or a species complex? Annu Rev Entomol 40:511–534

Brown JK, Idris AM, Alteri C, Stenger DC (2002) Emergence of a new cucurbit-infecting begomovirus capable of forming viable reassortants with related viruses in the *Squash leaf curl virus* cluster. Phytopathology 92:734–742

Candresse T, Cambra M (2006) Causal agent of sharka disease: historical perspective and current status of Plum pox virus strains. EPPO Bull 36:239–246

Celix A, Lopez-Sese A, Almarza N, Gomez-Guillamon ML, Rodriguez-Cerezo E (1996) Characterization of Cucurbit yellow stunting disorder virus, a *Bemisia tabaci*-transmitted closterovirus. Phytopathology 86:1370–1376

Cohen S, Antignus Y (1994) Tomato yellow leaf curl virus (TYLCV), a whitefly-borne geminivirus of tomatoes. In: Harris KF (ed) Advances in disease vector research, vol 10. Springer, New York, pp 259–288

Cui X, Li G, Wang D, Hu D, Zhou X (2005) A begomovirus DNAβ-encoded protein binds DNA, functions as a suppressor of RNA silencing, and targets the cell nucleus. J Virol 79:10764–10775

Daughtrey ML, Jones RK, Moyer JW, Daub ME, Baker JR (1997) Tospoviruses strike the greenhouse industry: INSV has become a major pathogen on flower crops. Plant Dis 81:1220–1230

Fauquet CM, Stanley J (2003) Geminivirus classification and nomenclature: progress and problems. Ann Appl Biol 142:165–189

Gao F, Bailes E, Robertson DL, Chen Y, Rodenburg CM, Michael SF, Cummins LB, Arthur LO, Peeters M, Shaw GM, Sharp PM, Hahn BH (1999) Origin of HIV-1 in the chimpanzee *Pan troglodytes troglodytes*. Nature 397:436–441

Garcia-Andres S, Accotto GP, Navas-Castillo J, Moriones E (2007) Founder effect, plant host, and recombination shape the emergent population of begomoviruses that cause the tomato yellow leaf curl disease in the Mediterranean basin. Virology 359:302–312

Garrido-Ramirez ER, Sudarshana MR, Gilbertson RL (2000) *Bean golden yellow mosaic virus* from Chiapas, Mexico: Characterization, pseudorecombination with other bean-infecting geminiviruses and germ plasm screening. Phytopathology 90:1224–1232

Gent DH, du Toit LJ, Fichtner SF, Mohan SK, Pappu HR, Schwartz HF (2006) Iris yellow spot virus: an emerging threat to onion bulb and seed production. Plant Dis 90:1468–1480

Gilbertson RL, Faria JC, Ahlquist PG, Maxwell DP (1993a) Genetic diversity in geminiviruses causing bean golden mosaic disease: the nucleotide sequence of the infectious cloned DNA components of a Brazilian isolate of bean golden mosaic virus. Phytopathology 83:709–715

Gilbertson RL, Hidayat SH, Paplomatas EJ, Rojas MR, Hou Y-M, Maxwell DP (1993b) Pseudorecombination between the cloned infectious DNA components of tomato mottle and bean dwarf mosaic geminiviruses. J Gen Virol 74:23–31

Guan Y, Zheng BJ, He YQ, Liu XL, Zhuang ZX, Cheung CL, Luo SW, Li PH, Zhang LJ, Guan YJ, Butt KM, Wong KL, Chan KW, Lim W, Shortridge KF, Yuen KY, Peiris JSM, Poon LLM (2003) Isolation and characterization of viruses related to the SARS coronovirus from animals in southern China. Science 302:276–278

Guzman P, Sudarshana MR, Seo Y-S, Rojas MR, Natwick E, Turini T, Mayberry K, Gilbertson RL (2000) A new bipartite geminivirus (begomovirus) causing leaf curl and crumpling in cucurbits in the Imperial Valley of California. Plant Dis 84:488

Hou Y-M, Gilbertson RL (1996) Increased pathogenicity in a pseudorecombinant bipartite geminivirus correlates with intermolecular recombination. J Virol 70:5430–5436

Idris AM, Brown JK (2004) *Cotton leaf crumple virus* is a distinct Western Hemisphere begomovirus species with complex evolutionary relationships indicative of recombination and reassortment. Phytopathology 94:1068–1074

James D, Glasa M (2006) Causal agent of sharka disease: new and emerging events associated with Plum pox virus characterization. EPPO Bull 36:247–250

Kon T, Sharma P, Ikegami M (2007) Suppressor of RNA silencing encoded by the monopartite tomato leaf curl Java begomovirus. Arch Virol 152:1273–1282

Kuo Y-W, Rojas MR, Gilbertson RL, Wintermantel WM (2007) First report of *Cucurbit yellow stunting disorder virus* in California and Arizona, in association with *Cucurbit leaf crumple virus* and *Squash leaf curl virus*. Plant Dis 91:330

Lanciotti RS, Roehrig JT, Deubel V, Smith J, Parker M, Steele K, Crise B, Volpe KE, Crabtree MB, Scherret JH, Hall RA, MacKensie JS, Cropp CB, Panigrahy B, Ostlund E, Schmitt B, Malkinson M, Banet C, Weissman J, Komar N, Savage HM, Stone W, McNamara T, Gubler D J (1999) Origin of the West Nile virus responsible for an outbreak of encephalitis in the northeastern United States. Science 286:2333–2337

Legg JP (1999) Emergence, spread and strategies for controlling the pandemic of cassava mosaic disease in east and central Africa. Crop Prot 18:627–637

Legg JP, Fauquet CM (2004) Cassava mosaic geminiviruses in Africa. Plant Mol Biol 56: 31–39

Leroy EM, Kumulungui B, Pourrut X, Rouquet P, Hassanin A, Yaba P, Delicat A, Paweska JT, Gonzalaz JP, Swanepoel R (2005) Fruit bats as reservoirs of Ebola virus. Nature 438:575–576

Li KS, Guan Y, Wang J, Smith GJD, Xu KM, Duan L, Rahardjo AP, Puthavathana P, Buranathai C, Nguyen TD, Estoepangestie ATS, Chaisingh A, Auewarakul P, Long HT, Hanh NTH, Webby RJ, Poon LLM, Chen H, Shortridge KF, Yuen KY, Webster RG, Peiris JSM (2004) Genesis of a highly pathogenic and potentially pandemic H5N1 influenza virus in eastern Asia. Nature 430:209–213

Mansoor S, Briddon RW, Zafar Y, Stanley J (2003) Geminivirus disease complexes: an emerging threat. Trends Plant Sci 8:128–134

Mendez-Lozano J, Torres-Pacheco I, Fauquet CM, Rivera-Bustamante RF (2003) Interactions between geminiviruses in a naturally occurring mixture: *Pepper huasteco virus* and *Pepper golden mosaic virus*. Phytopathology 93:270–277

Monci F, Sanchez-Campos S, Navas-Castillo J, Moriones E (2002) A natural recombinant between the geminiviruses Tomato yellow leaf curl Sardinia virus and Tomato yellow leaf curl virus exhibits a novel pathogenic phenotype and is becoming prevalent in Spanish populations. Virology 303:317–326

Marco CF, Aranda MA (2005) Genetic diversity of a natural population of *Cucurbit yellow stunting disorder virus*. J Gen Virol 86:815–822

Matson PA, Parton WJ, Power AG, Swift MJ (1997) Agricultural intensification and ecosystem properties. Science 277:504–509

Naidu RA, Deom CM, Sherwood JL (2005) Expansion of the host range of *Impatiens necrotic spot virus* to peppers. Plant Health Prog. doi:10.1094/PHP-2005-0727-01-HN

Navot N, Pichersky E, Zeidan M, Zamir D, Czosnek H (1991) Tomato yellow leaf curl virus: a whitefly-transmitted geminivirus with a single genomic component. Virology 185:151–161

Padidam M, Sawyer S, Fauquet CM (1999) Possible emergence of new geminiviruses by frequent recombination. Virology 265:218–225

Pagan I, Cordoba-Selles MDC, Martinez-Priego L, Fraile A, Malpica JM, Jorda C, Garcia-Arenal F (2006) Genetic structure of the population of *Pepino mosaic virus* infecting tomato crops in Spain. Phytopathology 96:274–279

Perring TM, Cooper AD, Kazmer DJ, Shields C, Shields J (1991) New strain of sweetpotato whitefly invades California vegetables. Calif Agric 45:10–12

Pita JS, Fondong VN, Sangare A, Otim-Nape GW, Ogwal S, Fauquet CM (2001) Recombination, pseudorecombination and synergism of geminiviruses are determinant keys to the epidemic of severe cassava mosaic disease in Uganda. J Gen Virol 82:655–665

Polston JE, Anderson PK (1997) The emergence of whitefly-transmitted geminiviruses in tomato in the Western Hemisphere. Plant Dis 81:1358–1369

Polston JE, McGovern RJ, Brown LG (1999) Introduction of tomato yellow leaf curl virus in Florida and implications for the spread of this and other geminiviruses of tomato. Plant Dis 83:984–988

Preiss W, Jeske H (2003) Multitasking in replication is common among geminiviruses. J Virol 77:2972–2980

Prins M, Goldbach R (1998) The emerging problem of tospovirus infection and nonconventional methods of control. Trends Microbiol 6:31–35

Roberts S, Stanley J (1994) Lethal mutations within the conserved stem-loop of African cassava mosaic virus DNA are rapidly corrected by genomic recombination. J Gen Virol 77:1947–1951

Rojas MR, Hagen C, Lucas WJ, Gilbertson RL (2005) Exploiting chinks in the plant's armor: evolution and emergence of geminiviruses. Annu Rev Phytopathol 43:361–394

Rojas MR, Kon T, Natwick E, Polston J, Fouad A, Gilbertson RL (2007) First report of *Tomato yellow leaf curl virus* associated with tomato yellow leaf curl disease in California, USA. Plant Dis 91:1056

Saeed M, Behjatnia SA, Mansoor S, Zafar Y, Hasnain S, Rezaian MA (2005) A single complementary-sense transcript of a geminiviral DNAβ satellite is determinant of pathogenicity. Mol Plant Microbe Interact 18:7–14

Salati R, Nahkla MK, Rojas MR, Guzman P, Jaquez J, Maxwell DP, Gilbertson RL (2002) *Tomato yellow leaf curl virus* in the Dominican Republic: characterization of an infectious clone, virus monitoring in whiteflies, and identification of reservoir hosts. Phytopathology 92:487–496

Saunders K, Salim N, Mali VR, Malathi VG, Briddon R, Markham PG, Stanley J (2002) Characterisation of Sri Lankan cassava mosaic virus and Indian cassava mosaic virus: evidence for acquisition of a DNA B component by a monopartite begomovirus. Virology 293:63–74

Saunders K, Norman A, Gucciardo S, Stanley J (2004) The DNA β satellite component associated with Ageratum yellow vein disease encodes an essential pathogenicity factor (βC1). Virology 324:37–47

Seal SE, van den Bosch F, Jeger MJ (2006) Factors influencing begomovirus evolution and their increasing global significance: implications for sustainable control. Crit Rev Plant Sci 25:23–46

Seo Y-S, Zhou Y-C, Turini TA, Cook CG, Gilbertson RL, Natwick ET (2006) Evaluation of cotton germplasm for resistance to the whitefly (*Bemisia tabaci*) and cotton leaf crumple (CLCr) disease and etiology of CLCr in the Imperial Valley of California. Plant Dis 90:877–884

Sharp LP, Hou Y-M, Garrido-Ramirez ER, Guzman P, Gilbertson RL (1999) A synergistic interaction between geminivirus DNA components results in increased symptom severity and viral DNA levels in plants. Phytopathology 89: S71

Simon AE, Roosinck MJ, Havelda Z (2004) Plant virus satellite and defective interfering RNAs: new paradigms for a new century. Annu Rev Phytopathol 42:415–437

Toscano NC, Castle SJ, Henneberry TJ, Prabhaker Castle N (1998) Persistent silverleaf whitefly exploits desert crop systems. Calif Agric 52:29–33

Towner JS, Khristova ML, Sealy TK, Vincent MJ, Erickson BR, Bawiec DA, Hartman AL, Comer JA, Zaki SR, Stroher U, Gomes da Silva F, del Castillo F, Rollin PE, Ksiazek TG, Nichol T (2006) Marburg virus genomics and association with a large hemorrhagic fever outbreak in Angola. J Virol 80:6497–6516

Vanitharani R, Chellappan P, Fauquet CM (2005) Geminiviruses and RNA silencing. Trends Plant Sci 10:144–151

Varma A, Malathi VG (2003) Emerging geminivirus problems: a serious threat to crop production. Ann Appl Biol 142:145–164

Verbeek M, Dullemans AM, van den Heuvel JFJM, Maris PC, and van der Vlugt RAA (2007) Identification and characterization of tomato torrado virus, a new plant picorna-like virus from tomato. Arch Virol 152:881–890

Verhoeven JTJ, van der Vlugt RAA, Roenhorst JW (2003) High similarity between tomato isolates of Pepino mosaic virus suggests a common origin. Eur J Plant Pathol 109:419–425

Whitfield A.E, Ullman DE, German TL (2005) Tospovirus-thrips interactions. Annu Rev Phytopathol 43:459–489

Wisler GC, Duffus JE, Liu H-Y, Li RH (1998) Ecology and epidemiology of whitefly-transmitted closteroviruses. Plant Dis 82:270–280

Chapter 4
Evolution of Integrated Plant Viruses

Thomas Hohn(✉), Katja R. Richert-Pöggeler, Christina Staginnus, Glyn Harper, Trude Schwarzacher, Chee How Teo, Pierre-Yves Teycheney, Marie-Line Iskra-Caruana, and Roger Hull

Abstract Plant pararetroviruses replicate their genome via a transcription–reverse transcription cycle like retroviruses, but unlike them their genomes do not obligatorily integrate into the host chromatin. Nevertheless, one can find complete or fragmented pararetrovirus genomes, as well as those from geminiviruses and even RNA viruses incorporated into the genomes of nearly all plants analysed. Integration events are thought to be rare and even rarer are those that find their way into the germ line. Normally, these integrated viral sequences are incomplete, rearranged and mutated and cannot easily escape as active viruses. However, in some cases apparently more recently acquired and therefore less initiated integrates can escape by direct transcription from tandem insertions or by recombination. This can lead to severe outbreaks in crop and ornamental plants. In anticipation of such events, methods have been developed for the detection and characterization of integrated virus sequences in plant genomes.

Thomas Hohn
Institute of Botany, University of Basel, Schönbeinstrasse 6, 4056 Basel, Switzerland
hohn@fmi.ch

M.J. Roossinck (ed.), *Plant Virus Evolution*.

4.1 Introduction

Some bacterial and animal viruses integrate their genomes into their host's chromatin as part of their replication cycle. Well-known examples include the lambdoid bacteriophages, but there are numerous others, including some animal DNA viruses, e.g. *Adeno associated virus* and polydnaviruses (PDVs) with double-stranded DNA (dsDNA) genomes, and the animal retroviruses with single-stranded (SS) RNA and double stranded (ds) DNA genomes, which are interconverted by viral reverse transcription and host transcription.

For integration, lambdoid bacteriophages code for a combined integrase/excise system (Mumm et al. 2006). Their proviral state is preserved by repression of the promoter and excision systems, but mobilization can be induced and leads to an autocatalytically controlled massive virus production.

Retrovirus genomes with terminal redundancy are integrated by the action of a virus-encoded integrase and can be mobilized by transcription starting and ending within the terminal repeats that contain promoter and polyadenylation signals. For retroviruses the integrated state is also preserved by inhibition of transcription and RNA processing (Coffin et al. 1997). However, once induced, again virus replication is autocatalytically accelerated. Viruses thus escape host detection in the proviral phase and out-titrate the defence mechanism in the productive phase.

Transposable elements are also parasitic, integrating elements found in all kingdoms. There are two types of transposons. DNA transposons leave the chromatin by excision and reinsert themselves at a new location under the action of their own transposase or one provided by a helper element. Retrotransposons integrate as DNA and are mobilized by transcription much like retroviruses, but plant retrotransposons usually do not move from cell to cell or organism to organism. The presence of a retrovirus-like envelope (env) gene necessary for viral infectivity in some members of long terminal repeat (LTR) retrotransposons (Laten et al. 1998) and the ability to form viruslike particles led to the classification of *Ty3-gypsy* retroelements as *Metaviridae* and the *Ty1-copia* elements as *Pseudoviridae* (Fig. 4.1).

Some viruses do not encode integrases but still integrate their genome into the genome of their hosts. PDVs integrate randomly into chromosomal DNA of their wasp hosts that parasitize other insects (Kroemer and Webb 2004). PDVs persist in the genome of associated wasps as stably integrated proviruses. They replicate in the ovaries of female wasps, who subsequently inject infected eggs and virions into their Lepidoptera hosts' larval stage. Expression of PDV genes in Lepidoptera is essential for survival of the parasitoid's offspring (Schmidt et al. 2001). Sequencing of integration loci of PDVs in the genome of their wasp hosts suggests a host-mediated transposition mechanism for viral DNA insertion (Gundersen-Rindal and Lynn 2003).

Spontaneous integration of viral sequences into the host cell DNA also has been reported for viruses associated with the development of various cancers (Ferber et al. 2003), such as hepadnaviruses (HBVs), animal- and human-infecting pararetroviruses (Yang and Summers 1999), adenoviruses (Orend et al. 1994) and human

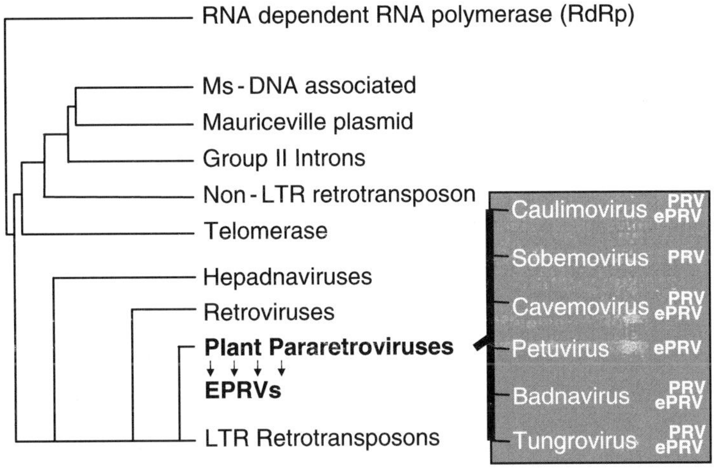

Fig. 4.1 Phylogeny of elements containing reverse transcriptase and relationship to RNA-dependent RNA polymerase. *LTR* long tandem repeat, *EPRV* endogenous pararetrovirus, *msDNA* multicopy single-stranded DNA

papillomavirus (HPV; Wentzensen et al. 2004). Some hepadnaviral insertions activate members of the *myc* family of protooncogenes in hepatocellular carcinomas of animals but not in those of humans (Tsuei et al. 2002; Bill and Summers 2004). However, the impact of HBV and HPV genome invasion on oncogenesis still requires further investigation (Ferber et al. 2003). No activation of HBV or HPV proviral sequences has been reported to date.

4.2 Plant DNA Viruses

Three families of plant viruses, the *Geminiviridae*, the *Nanoviridae* and the *Caulimoviridae*, have DNA genomes. The four genera of geminiviruses (*Begomovirus*, *Curtovirus*, *Mastrevirus* and *Topocuvirus*) are classified according to their insect vector(s), host range and genome organization (Fauquet et al. 2005). Their single-ssDNA genomes consist of one or two components. They are converted to dsDNA when they enter the nucleus and replicate by a rolling-circle mechanism (Fig. 4.2).

The *Nanoviridae* are divided into two genera (*Nanovirus* and *Babuvirus*). Their integral genome consists of up to 12 ssDNAs. Their number and type have not been experimentally determined yet for any of the species. All of them seem to be positive sense, transcribed in one direction and containing a conserved stem loop structure in the noncoding region (Fauquet et al. 2005).

The *Caulimoviridae* or plant pararetroviruses comprise six genera of plant pararetroviruses, the icosahedral *Caulimovirus*, *Sobemovirus*, *Cavemovirus* and

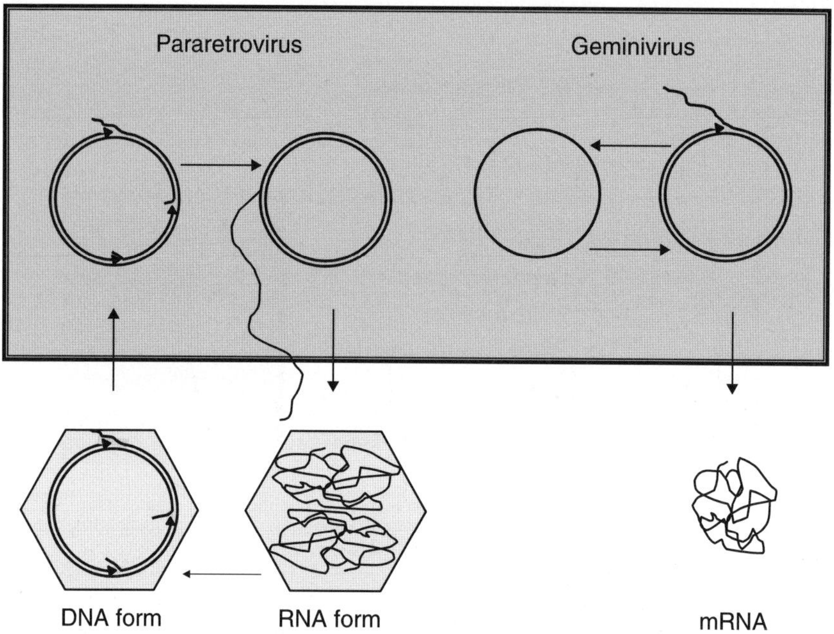

Fig. 4.2 Replication of pararetroviruses and geminiviruses. In both cases, double-stranded DNAs (in the form of minichromosomes) accumulate in the nucleus and are transcribed. Geminivirus DNA is amplified via a rolling circle in the nucleus and pararetrovirus DNA by reverse transcription of the terminally redundant pregenomic RNA in the cytoplasm. *mRNA* messenger RNA

Petuvirus and the bacilliform *Badnavirus* and *Tungrovirus*, which are defined by the number and order of open reading frames (ORF) of their genomes (Fauquet et al. 2005; Fig. 4.1). The close relationship between retroviruses and some LTR transposons, in particular the *Ty3-gypsy Metaviridae* elements, is evidenced by the same order of *pol* gene domains: protease–reverse transcriptase–ribonuclease H (RH) gene followed by the intergrase (IN) gene (Hansen and Heslop-Harrison 2004). The same order holds for the plant pararetroviruses, albeit the integrase domain is missing. On the DNA level, reverse transcriptase domain sequences of *Metaviridae* elements and retroviruses are very similar (Xiong and Eickbush 1990), but still show significant variation and base pair changes that do not always allow a clear distinction. On the amino acid level endogenous pararetroviruses (EPRVs) and *Metaviridae* are extremely conserved in the reverse transcriptase domain, while the RNase H domain contains several significant motifs (Teo and Schwarzacher, unpublished data) that can be used to design primers to distinguish plant pararetroviruses from retrotransposons (Richert-Pöggeler and Shepherd 1997; Hansen and Heslop-Harrison 2004). A maximum-likelihood tree based on the amino acid sequences, including one of the specific boxes, separates pararetroviruses from retroelements unanimously and groups them into several significant clades (Fig. 4.3).

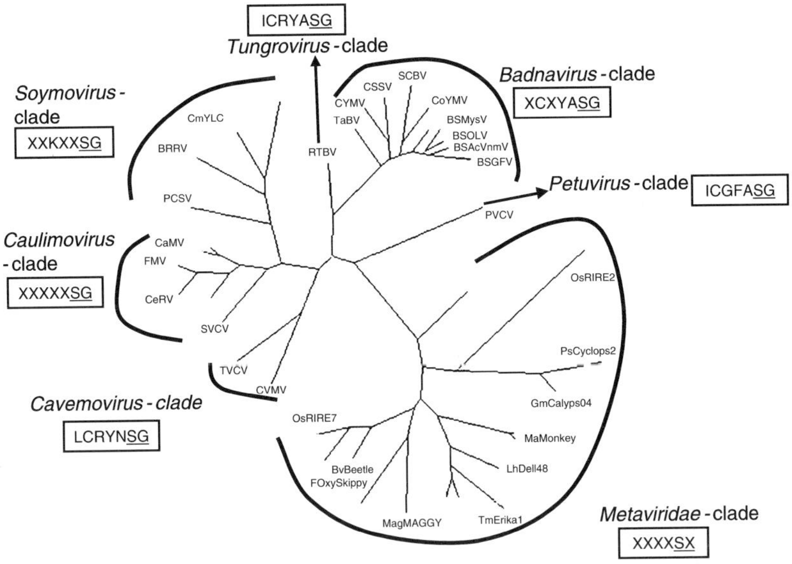

Fig. 4.3 Maximum-likelihood tree based on comparisons of amino acid sequences of the reverse transcriptase and RNase H domains of EPRVs and *Metaviridae*. Phylogenetic analysis of 15 *Metaviridae* retrotransposon sequences and 24 pararetrovirus reverse transcriptase/RNase H domains. The phylogenetic tree was constructed based on the Jones–Taylor–Thornton model (Jones et al. 1992). The conserved motif in the RNase H domain at position 365 with its clade-specific variants is provided in the *boxes*; two amino acid residues separate the *Metaviridae* retrotransposons and pararetroviruses (*underlined*)

As with the animal retroviruses, plant pararetroviruses replicate via a transcription–reverse transcription process, i.e. they produce a terminally redundant RNA form of their nuclear genome, transport it to the cytoplasm, incorporate two copies of it into viral particles and reverse transcribe them. Viral particles dock to the nuclear pore and release the open circular DNA into the nucleus, where it is converted into supercoiled DNA and complexed with histones to form a minichromosome (reviewed in Hull 2001; Hohn and Richert-Pöggeler 2006; Fig. 4.2).

4.3 Detection of Integrated Plant DNA Virus Sequences

The genomes of plant DNA viruses and pararetroviruses do not actively integrate into their host's chromatin, as integration is not required for viral replication. Viral genomes existing as minichromosomes in the nucleus of infected cells do not code for enzymes for integration and excision, nor do they have the inverted or direct terminal repeats recognized by those enzymes. Nevertheless, integrated geminivirus- and pararetrovirus-related sequences have been detected in the genomes of several plant species and have been termed geminivirus-related DNA (GRD; Bejararno et al. 1996) and EPRVs (Matzke et al. 2000; Harper et al. 2002; Staginnus and Richert-Pöggeler 2006), respectively. The latter are designated with the prefix "e" in front of the virus acronym.

Several observations led to the discoveries of geminiviral and pararetroviral sequences within plant genomes. In the case of geminiviruses, Day et al. (1991) constructed *Tomato golden mosaic virus* (TGMV) resistant tobacco plants expressing TGMV antisense RNA and unexpectedly found that genomic DNA of untransformed control plants cross-hybridized with a TGMV AC1 probe (for genome organization of begomoviruses, see Hull 2001). A systematic analysis of a tobacco (*Nicotiana tabacum* 'Samsun') DNA library led to the characterization of several AC1 GRDs (later called the GRD5 family) present in multiple repeats. On the basis of Southern blot analysis their total number was estimated to be around 360, all clustered within a single 340-kb restriction fragment (Bejarano et al. 1996). Sequence comparisons showed that these GRDs were most similar to New World begomoviruses. Interestingly, sequences hybridizing to AC2, AC3 or AV1 were not detected. The AC1 sequences contain both *cis* and *trans* elements required for replication, i.e. the geminiviral replication origin plus the adjacent *Rep* gene. This property might have allowed them to become amplified as a type of "intrachromosomal replicon", but they are not known to function as extrachromosomal elements as they are incomplete viruses. Further analysis showed that GRD5 family members are present in *N. kawakamii*, *N. tomentosa* and *N. tomentosiformis* but not in nine other more distantly related *Nicotiana* species (Ashby et al. 1997). A second family, GRD3, was found only in *N. tomentosiformis* and its integration is therefore thought to have occurred more recently than that of GRD5 (Murad et al. 2004). In the case of pararetroviruses, Lafleur et al. (1996) found integrated viruslike sequences (i.e. *Banana streak virus*; BSV) in certain *Musa* genomes by Southern blots.

The first indications of activatable EPRVs came from observations of apparently spontaneous viral infections in petunia, tobacco and banana by *Petunia vein clearing virus* (PVCV; Richert-Pöggeler and Shepherd 1997; Richert-Pöggeler et al. 2003), *Tobacco vein clearing virus* (TVCV; Lockhart et al. 2000) and BSV (Ndowora et al. 1999; Harper et al. 1999; Dallot et al. 2001), following stress, wounding and tissue culture in insect-free environments, respectively, suggesting that they were derived from integrated forms.

Subsequently, other pararetroviral sequences integrated into the plant genome were characterized. They belong to four of the six pararetrovirus genera, i.e. *Petuvirus*, *Cavemovirus*, *Tungrovirus* and *Badnavirus* (Figs. 4.1, 4.3). Integrated forms of *Caulimovirus* and *Soymovirus* have not been found in the genome of plants yet. Evidence for possibly more ancient and mostly noninducible EPRVs came by serendipity during sequencing of certain silenced loci of the tobacco genome. With use of information from the endogenous TVCV (eTVCV) sequence, further independent and incomplete inserts were found and a virtual *Tobacco endogenous pararetrovirus* (NsEPRV, formerly designated TEPRV) genome was assembled (Jakowitsch et al. 1999, Matzke et al. 2004).

Methods for recognizing EPRVs based on known sequences include Southern hybridization of total genomic DNA, analysis of DNA libraries by hybridization and PCR using pairs of consensus sequences as primers. These approaches have led to the detection of a number of new EPRV sequences in potato (Hansen et al. 2005), tomato (Staginnus et al. 2007) and a diverse collection in *Musa* species (Geering et al. 2005a, b). While the amino acid sequence is highly conserved, at the DNA sequence level much greater degeneracy is found, such that a universal PCR approach has been difficult (Schwarzacher and Kubis, unpublished data). However, using a double-PCR method with nested and seminested primers, EPRV sequences from diverse plant species were isolated, including Norway spruce, pine, *Brassica*, olive, barley and sugar beet (Paradigm consortium, unpublished data). This further confirms the indications for a wealth of additional EPRV sequences in the whole plant world (Hansen 2003; Staginnus and Richert-Pöggeler 2006).

The ultimate tools for the identifications of EPRVs are genomic sequencing or data mining of the published sequenced genomes. At the time of writing this review, three plant genomes have been completely sequenced. While EPRVs could not be found in the genome of *Arabidopsis thaliana*, a full screening of the recently published genome of black cottonwood (*Populus trichocarpa* Torr. & Gray; Tuskan et al. 2006) has not yet been completed. Meanwhile, a systematic search of the rice (*Oryza sativa* 'Nipponbare') genome revealed *Rice tungro bacilliform virus* (RTBV) related sequences (Kunii et al. 2004).

4.4 EPRVs Identified in Plant Genomes

Distinct BSV-EPRVs have been described for different *Musa* species (Harper et al. 1999; Ndowora et al. 1999; Geering et al. 2001, 2005; Safár et al. 2004; Harper et al. 2005; Table 4.1). Substantial evidence suggests that Obino l'Ewai (BSOLV)-EPRVs are activatable in interspecific *Musa* hybrids (Ndowora et al. 1999; Harper et al. 1999; Lheureux et al. 2003; Dallot et al. 2000); integrants of BSV-related species like Goldfinger (BSGFV) or Imove (BSImV) were suspected to cause infection (Safár et al. 2004; Geering et al. 2005b).

Four different BSV species (BSOLV, BSGFV, BSMysV and BSImV), easily distinguishable from each other with less than 85% sequence similarity, are found as both episomal and integrated virus sequences in *Musa balbisiana* (Pifanelli et al. 2005) or *Musa* hybrids. Other BSV isolates belonging to the clade that includes these four species originated from natural epidemics and were recently identified and proposed to arise from vertical transmission of activatable integrated sequences within B genome bananas (Harper et al. 2005; Fargette et al. 2006), although no matching EPRV have been identified so far in the B genome (Teycheney, unpublished data). Furthermore, numerous badnavirus-related banana endogenous viruses (BEVs) were found in *Musa* (Geering et al. 2005b). The very high rate of molecular diversity displayed by BEVs tends to show that they stem from multiple independent integration events, and that integration of pararetroviral sequences into

Table 4.1 The best studied complete and incomplete endogenous viral sequences in plants

Family	Genus	Name	Sequences	Inducible	Host	Comments	Recent reference
Viroid		Carnation small viroid like (CarSV) RNA	Complete	Yes	Carnation; *Carnation etched ring virus*		Hegedüs et al. (2004)
Potyviridae		*Potyvirus*-like	Coat protein	No	*Vitis vinifera*	With MITE-type retroelement	Tanne and Sela (2005)
Geminiviridae	*Begomovirus*	*Geminivirus*-related DNA (GRD) 5	AC1 only	No	*Nicotiana tomentosiformis*; *N. tomentosa*; *N. kawikamii*		Ashby et al. (1997)
Geminiviridae	*Begomovirus*	GRD 3	AC1 only	No	*N. tomentosiformis*		Murad et al. (2004)
Caulimoviridae	*Petuvirus*	*Petunia vein clearing virus* (PVCV)	Complete	Yes	*Petunia hybrida*		Richert-Poeggeler et al. (2003)
Caulimoviridae	*Cavemovirus*	*Tobacco vein clearing virus* (TVCV)	Complete	Yes	*N. edwarsonii*		Lockhart et al. (2000)
Caulimoviridae	*Cavemovirus*	TVCV-like (NtoEPRV; NsEPRV)	Complete	No	*N. tabacum*; *N. sylvestris*		Mette et al. (2002)
Caulimoviridae	*Cavemovirus*	TVCV-like (LycEPRV)	Complete	No	*Nicotiana* (various species)		Staginnus et al. (2007)
Caulimoviridae	*Cavemovirus*	TVCV-like (SoTu)	TAV, part IGR	No	*Solanum lycopersicon* (various species)	3 families	Hansen et al. (2005)
Caulimoviridae	*Tungrovirus*	*Rice tungro bacilliform virus* (RTBV)	Without ORF2	No	*Oryza sativa*	3 phylogenetic clusters	Kunii et al. (2004)

Caulimoviridae	*Cavemovirus*	*Dahlia mosaicvirus*	Complete	Yes	*Dahlia punata*		Pahalawatta et al. (2007)
Caulimoviridae	*Badnavirus*	Banana endogenous viruses (BEV)	Reverse transcriptase/ RNase region of ORF3	No	*Musa balbisana; M. accuminata*		Geering et al. (2005b)
Pararetrovirus	*Badnavirus*	BSV Obino l'Ewai (BSOLV)	Complete	Yes	*M. balbisana*	2 chromosomal locations	Harper et al. (1999)
Pararetrovirus	*Badnavirus*	BSV (BSMysV)	Complete	No	*M. balbisiana*		Geering et al. (2005a)
Pararetrovirus	*Badnavirus*	BSV Imove (BSImV)	Complete	Yes?	*M. balbisana*		Unpublished
Pararetrovirus	*Badnavirus*	BSV Goldfinger (BSGfV)	Complete	Yes?	*M. balbisana*		Unpublished

the *Musa* genome has occurred frequently. However, it should be noted that banana is usually propagated vegetatively using suckers and only rarely from seed.

Endogenous forms of either *Petuvirus* or *Cavemovirus* have been isolated from *Solanaceae* species. Endogenous PVCV (ePVCV) has been detected in several *Petunia* species and hybrids, but is absent from the *P. parodii* genome (Harper et al. 2002, Richert-Pöggeler et al. 2003). Integrated copies can cause infection under certain conditions (Richert-Pöggeler et al. 2003), commonly in hybrids and only very occasionally in the parental *P. axillaris*. Interestingly, no virus vector for horizontal transmission of ePVCV or eTVCV has been identified yet, and survival of the virus population is only guaranteed by vertical transmission.

The eTVCV from *N. edwardsonii* most likely gives rise to episomal infectious particles, but so far only vertical and no horizontal transmission has been reported (Lockhart et al. 2000). Factors that initiate activation are discussed in Sec. 4.6.8.

In addition to the activatable EPRV families in *Solanaceae* and *Musaceae*, further nonactivatable pararetroviral integrants have been detected in various monocotyledenous and dicotyledenous species (Table 4.1). In the *Solanaceae* EPRVs, e.g. Nicotiana Sylvestis (Ns) EPRV or (Nto) EPRV in *Nicotiana* (Jakowitsch et al. 1999; Gregor et al. 2004), SoTu in *Solanum tuberosum* (Hansen et al. 2005) and *Lycopersicon* (Lyc) EPRV in (Staginnus et al. 2007), form a TVCV-like group. However, viruslike sequences could only be assembled from various defective genomic copies and no corresponding episomal form has been detected. The RTBV sequences (eRTBV) in the rice genome fall into three distinct phylogenic clusters (Kunii et al. 2004). The genomes of putative RTBVs were segmented in the genome and lacked ORF2, and thus were considered noninfectious. Even in lower plants and gymnosperms, the existence of similar related sequences could be detected by hybridization and PCR (Hansen 2003).

The ubiquitous appearance of EPRVs suggests that the integration of pararetroviral sequences is a common feature within in the plant kingdom. However, some members of the *Caulimoviridae* and *Geminiviridae* families have wide host ranges but have not (yet) been detected as integrated sequences. This raises various questions about viruses with a wide host range such as: Are they more controlled within the cell such that replication intermediates are excluded from the host genome? Do they possess genes that allow wider infectivity but reduce the opportunity for integration? Do they have subtly different mechanisms of replication?

4.5 Integration of DNA Copies of RNA Viruses and Viroids

DNA copies of coat protein sequences of an RNA virus, *Potato virus Y*, have been found integrated into the genome of grapevine (Tanne and Sela 2005) and viroid-like sequences have been found in carnation (Hegedüs et al. 2004). Reverse transcription must have been involved in these integration events. In fact, the potyvirus sequence was connected to a MITE-type retroelement, and the viroid sequence was found incorporated into the genome of *Carnation etched ring caulimovirus* (Vera et al. 2000). Substantial natural integration of virus RNA causing resistance to cognate *Dicistroviruses* was recently shown for the bee genome (Maori et al. 2007).

4.6 Evolution

The study of the evolution of integrated viral sequences presents several specific difficulties. To reconstruct the steps of evolution, experimental approaches are limited since only selected examples can be investigated thoroughly and therefore may not be representative of all integrated sequences. The evolution of viruses and their integrants cannot be assessed by the standard "stratigraphic" approaches used for higher organisms but can be by "molecular fossils" (Lovisolo et al. 2003). Identification of remnants of viral sequences in DNA databases of plants and other organisms reveals such "molecular fossils" and could help to put the palaeovirologist in a similar situation to that of the palaeontologist, collecting evolution data on a stratigraphic basis.

There are various selection pressures acting on the stages of integration and maintenance of viral sequences in the host genome, and these are discussed in the following sections.

4.6.1 Prehistory of EPRVs and GRDs

A diversity of related elements in eukaryotes and prokaryotes use reverse transcription as a mode of replication (Eickbush 1997; Fig. 4.1). Reverse transcriptase is a very ancient enzyme and is thought to be derived from the even older RNA-dependent RNA polymerase. One branch of these elements comprises LTR transposons, retroviruses and pararetroviruses. Such elements were likely built from modules (Botstein 1980; Hull 1992, 2001; Hull and Covey 1996; see Chap. 8). The core module consists of reverse transcriptase and a capsid protein protecting the RNA form of the element and containing the reverse transcription reaction. Further modules determine other properties of the elements: an integrase function is used by retroviruses and retrotransposons for incorporation into the host chromatin; an envelope protein as part of a lipid membrane is used by animal retroviruses to invade new host cells; and movement proteins and insect transmission factors are used by plant pararetroviruses to move from cell to cell through modified plasmodesmata and to specifically interact with insect vectors. These modules might have been acquired and exchanged from other elements and viruses: integrase from DNA transposons, envelope protein from other animal viruses and movement proteins and insect transmission factors from other plant viruses. Alternatively, as suggested by Hull (1992, 2001) and Hull and Covey (1996), they may have been acquired directly from their host and subsequently modified.

Further accessory proteins have been acquired that control various steps of gene expression and host interaction.

On the other hand, the loss of modules from a virus may also have contributed to the diversity of today's endogenous retroelements. *Diaspora*, a member of the *Metaviridae*, identified in *Glycine max* is assumed to originate from a retrovirus lineage by loss of the envelope gene (Yano et al. 2005). This loss might have occurred during adaptation to a heterologous, e.g. plant host, since plant viruses usually cannot make use of envelope proteins to enter a plant cell. Another example of possible loss of a module is PVCV, which lacks an aphid transmission factor typical of caulimoviruses in both its episomal and its chromosomal forms (Richert-Pöggeler and Shepherd 1997).

Similarly, geminiviruses may be built from modules shared with animal and plant viruses. For example, they share with animal parvoviruses the ssDNA genome as well as the replication strategy and with papovaviruses the bidirectional transcription strategy and a replication protein (AC1; T-antigen, respectively) interacting with the origin of replication, nicking dsDNA, sequestering the host DNA polymerase and manipulating the cell cycle (Gutierrez 2000). Again their movement proteins are related to those of other plant viruses.

A feature common to all viruses is that they condense a large amount of information in little space. Furthermore, they are infectious entities. Such characteristics make them ideal vehicles for horizontal gene transfer contributing to nature's diversity (Maori et al. 2007). Although unable to replicate without the host metabolic machinery, they can provide powerful elements for driving gene expression and, once incorporated into the plant genome, can exert a force on host evolution.

The driving forces for passive EPRV and GRD integration into specific sites might involve both the virus and the plant host and include the greater genome plasticity in plants when compared with the highly conserved genomes of mammals (Bennetzen 2005).

4.6.2 Integration Mechanisms

Any DNA that gets into the nucleus has a chance, albeit low, of becoming integrated into the chromatin. This fact is experimentally well established on the basis of the transformation of plant cells by direct gene transfer (Paszkowski et al. 1984). The generally accepted mechanism of such integration events is double-strand break repair (Puchta 2005), which can lead to either homologous recombination or random integration, the latter being the predominant mode in plants. Two mechanisms are proposed for double-strand break repair; synthesis-dependent strand annealing and single-strand annealing (Puchta 2005). They both involve ssDNA ends, which either invade DNA D-loops or bind to microhomologies at breaks of the acceptor DNA. Double-strand break repair can lead to mitotic homologous recombination, if donor and acceptor share high sequence identity. However, in plants, homologous recombination between host and invading DNAs is very rare, probably because the mechanism of scanning the chromatin for similarities is inefficient. More frequently, short "microhomologies" might be involved in initiating the invading single strand. Base-pairing within AU-rich regions has been assumed to facilitate template switches from viral to host RNAs in other viruses (White and Nagy 2004; see Chap. 7).

Complete or partial ssDNA, readily available from both plant pararetroviruses and geminiviruses, is an ideal primer and template to initiate the recombination process (Fig. 4.2). During reverse transcription of pararetroviral pregenomic RNA, an RNA/DNA hybrid is first synthesized, followed by degradation of the RNA template by the viral RNase H, leading to ssDNA. Then the second DNA strand is produced using this ssDNA as a template and the remaining RNA fragments as primers. Thus, there are periods between RNase H action and the production of the second DNA strand when ssDNA is present and it can remain so if second-strand production is incomplete. The final reverse transcription products are double-stranded open circular DNA forms with single-stranded overhanging sequences (flaps). After they enter the nucleus, these forms are "repaired" by removing the flaps, filling gaps and ligation of the ends to yield supercoiled circular DNA. While the ssDNA reverse transcription intermediates might not be available to initiate recombination, because the reverse transcription process usually occurs in the cytoplasm, viral DNA molecules with single-strand flaps are at least transiently available in the nucleus and these flaps form perfect sites for single-strand invasion (Fig. 4.4). For ePVCV (Richert-Pöggeler et al.

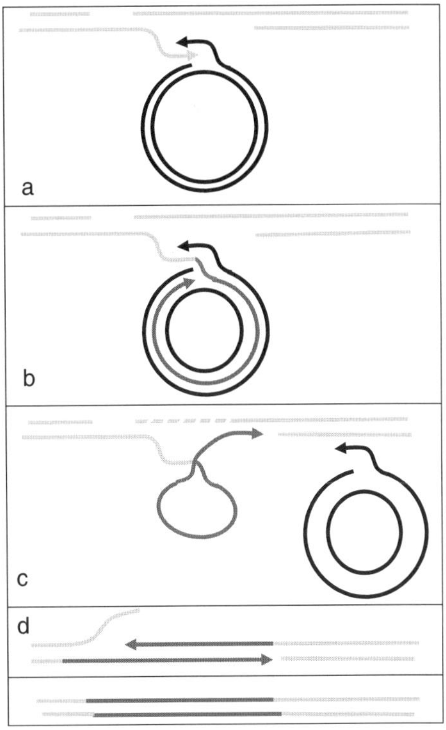

Fig. 4.4 Double-strand break repair involving single-strand invasion. Following a chromosomal break, single-stranded (ssDNA) can invade a rolling-circle replicative intermediate of a geminivirus DNA or an open circular form of a pararetrovirus DNA (**a**) and elongate from the viral template (**b**). Microhomologies at the DNA break allow reannealing of the hybrid DNA strand (**c**) and restoration of the chromosome by filling and ligation (**d**)

2003), NsEPRV (Jakowitsch et al. 1999) and eRTBV (Kunii et al. 2004), flap regions were found to be integration sites.

An alternative mode of integration for pararetroviruses might be by hitchhiking on retrotransposons. Pararetroviruses and retroviruses have a high recombination rate (Hohn 1994; Froissart 2005). Template switches occur obligatorily and also illegitimately between viral RNA strands during reverse transcription (Hohn 1994). Foreign RNA, such as host messenger RNA, could be incorporated into pararetrovirus particles, combined by illegitimate replicative recombination and lead to hybrid DNA molecules (Mayo and Jolly 1991).

Retrotransposon/pararetrovirus hybrids capable of true transposition could be created by this route, too. Indeed, EPRVs found in the neighbourhood of *Metaviridae* (*Ty3-gypsy* elements; Richert-Pöggeler et al. 2003; Gregor et al. 2004, Staginnus et al. 2007) could be products of such events. However such a neighbourhood could also be explained by a common preference for certain chromosomal locations as integration sites.

Geminiviruses are replicated in the nucleus by a rolling-circle mechanism, creating linear ssDNA intermediates, which are eventually circularized. Such single-strand intermediates are indeed very recombinogenic and have been shown to initiate a recombination-dependent replication mechanism (Alberter et al. 2005). Eukaryotic transposons called "helitrons" that transpose by rolling-circle replication have been suggested as the origin of geminiviruses (Kapitonov and Jurka 2001).

4.6.3 Clustering

Clustering of a few to hundreds of EPRV copies has been observed in many cases (Jakowitsch et al. 1999). Those clusters become obvious, particularly in solanecous species, by fluorescent in situ hybridization (FISH) experiments with prominent signals found on several chromosomes (Richert-Pöggeler et al. 2003; Fig. 4.5a), and by sequencing long EPRV-containing DNA fragments (Staginnus et al. 2007; Gregor et al 2004). In *Musa*, BSV FISH signals are very weak, indicating that single or only a few copies have been integrated (Schwarzacher, Teo and Iskra-Caruana, unpublished data; Fig. 4.5b, c). This is also supported by copy number estimates from a bacterial artificial chromosome sequencing project (Pifanelli and Iskra-Caruana, unpublished data).

Several mechanisms could be responsible for clustering. Clusters can form when several copies of an EPRV are involved simultaneously in a single break-repair event. Such clusterings have been observed during *Agrobacterium tumefaciens* T-DNA integration and integration of transgenes (Kononov et al. 1997; Ohba et al. 1995; Hanin and Paszkowski 2003).

For at least some bacilliform pararetroviruses, encapsidated concatamers of viral genomes that could have been generated during replication have been found (Geijskes et al. 2004). Furthermore, extrachromosomal recombination of foreign

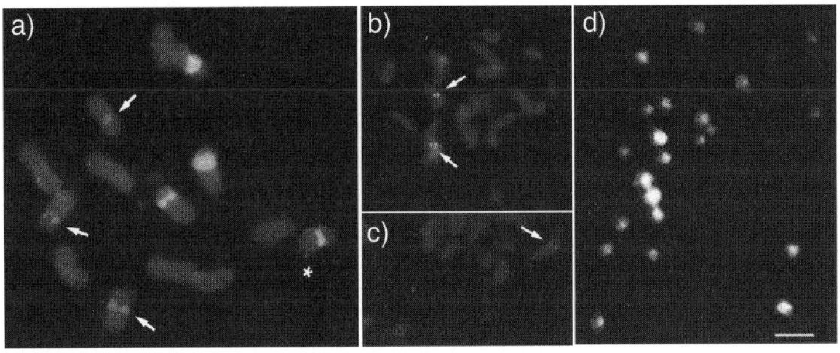

Fig. 4.5 Fluorescent in situ hybridization of EPRV and related sequences in banana and petunia. **a** *Petunia hybrida* 'V26' (2*n*=14) chromosomes. *Petunia vein clearing virus* sequences (*red signal*) form large clusters near the centromeres of three chromosomes (*arrows*) with minor sites at intercalary or telomeric sites (*asterisk*); 5S ribosomal DNA signal is shown in *cyan*) chromosomes counterstained with 4′,6-diamidino-2-phenylindole (DAPI; *blue*) **b–d** *Musa balbisiana* (B genome, 2*n*=22) chromosomes (stained blue with DAPI). Two BSOLV (**b**, *cyan*) and one BSVImV (**c**, *red*) integration sites (*arrows*) were detected. Pseudovirdae sequences (*red*) and the BSOLV flanking repeats (Harper et al. 1999; *cyan*) are localized at the centromeres of all chromosomes (appearing *yellow-white* where overlapping in **d**)

DNA after uptake has been shown to occur in plants (reviewed in Hanin and Paszkowski 2003). These observations indicate that arrays or clusters of viral sequences could have been formed prior to genome invasion.

If at least two copies are integrated, sister-chromatin exchange can lead to higher-order clustering (Wilson and Thompson 2007). Very large clusters of hundreds of units that have been observed in some cases could have arisen from several successive cycles of sister-chromatin exchange. Although homologous integration is rare in plants, these rare cases should still be considered as possible causes of clustering by recombination of episomal viral genomes with integrated forms.

Finally clustering might be caused by repetitive integration of choosy pararetroviruses preferring certain chromosomal regions for the integration event. AT-rich target sequences have been suspected to attract RTBV integrants (Kuuni et al. 2004).

4.6.4 Sites of EPRV Integration

EPRVs are preferentially found integrated in heterochromatin, mainly in the pericentromeric regions, but also in some intercalary regions and more rarely at the telomere (Richert-Pöggeler et al. 2003; Hansen et al. 2005; Staginnus et al. 2007, Schwarzacher and Richert-Pöggeler, unpublished data; Fig. 4.5a–c). Pericentromeric regions are also preferentially inhabited by *Metaviridae* and *Pseudoviridae* sequences in most plant species, including those where EPRVs have been found (Kumar and Bennetzen 1999; Wang et al. 2006). The physical colocalization of EPRVs and retroelement sequences is evidenced by FISH experiments with EPRV flanking sequences in banana (Fig. 4.5d) and by identifying EPRV adjacent sequences in genomic clones (Richert-Pöggeler et al 2003; Staginnus et al. 2007; Teo and Schwarzacher, unpublished data).

It is still unknown if the favoured EPRV incorporations into the pericentromeric regions are governed by intrinsic integration preferences, as was suggested for transposons by Malik and Eikbush (1999) (see also Berry et al. 2006), or are the result of EPRVs/host chromatin coevolution. EPRVs incorporated into euchromatin are more likely to interfere with expression of important genes and to reduce plant fitness. As a consequence there would be selection pressure either against the affected plants or in favour of EPRV removal. On the other hand, EPRVs in a heterochromatic environment may not dramatically affect plant fitness and hence may be retained. Furthermore, the expression of viral genes detrimental to the host would likely be suppressed in heterochromatin. Similarly, recent evidence indicated that the clustering of transposons in *A. thaliana* is governed by purifying selection (Kato et al. 2004).

Some plant genomes harbour many different but related groups of integrated EPRVs, with different copy numbers and in different stages of degeneration. In *Musa*, both potentially activatable and degenerate elements have been identified, and sequence comparisons suggest several independent integration events (Geering et al. 2005b; Lheureux et al. 2007). Despite the fact that different groups of integrants might each originate from a cognate exogenous pararetrovirus, each genome seems to harbour only EPRVs of a single *Caulimoviridae* genus. Variation in EPRV copy numbers between different species might reflect the number of invasion events (Matzke et al. 2004), or subsequent phases of amplification or elimination from the host genome, as reported for retrotransposons and the endogenous forms of retroviruses (Kalendar et al. 2000; Gifford and Tristem 2003; Skalická et al. 2005).

4.6.5 Entering the Germ Line

Somatic integration of viral sequences might occur relatively frequently, but it is only very rarely sexually transmitted. Two routes might be envisaged for viral sequences to enter the germ line. Viruses generally do not enter meristems, perhaps because host defence mechanisms such as gene silencing are most active in that tissue. This allows recovery from infection in newly formed tissue and prevents seed transmission. As pointed out by Schwach et al. (2005), true meristem exclusion is restricted to the growing point and does not include recovery from virus infection in the uppermost leaves. For some but not all viral infections the host RNA-dependent RNA polymerase RDR6 seems to play an important role in protecting the growing point (Schwach et al. 2005; Blevins et al. 2006). Only selected phloem-transported endogenous RNA molecules can enter the shoot apex (Foster et al. 2002). Intercellular movement of DNA viruses is connected with tubule formation (Hull 2002) along plasmodesmata that are missing in meristematic cells. It is possible that occasionally viral nucleic acids are overlooked by the meristematic surveillance system, providing DNA molecules for integration that are not abundant or complete enough to cause infection of this tissue.

Alternatively, a somatic cell carrying an EPRV could participate in the de novo organization of a meristem and the formation of a scion during vegetative propagation either naturally or by tissue culture. Progenies of these cells must finally get into the L2 layer, involved in gametophyte genesis (Steeves and Sussex 1989). If the presence of an EPRV conferred a selective advantage, then the rare events of the virus entering the meristem would be selected in the long term.

4.6.6 Fate of EPRVs

Pararetroviruses show striking evolutionary similarities to the endogenous forms of retroviruses (Gifford and Tristem 2003; Staginnus and Richert-Pöggeler 2006), probably owing to common regulatory pathways of the host genome during colonization. One can imagine an initial genome invasion event being followed by amplification of the element and the host's response for EPRVs, as discussed in Sec. 4.6.3. This would lead to mutation and/or to loss of copies. Transcriptional gene silencing (TGS) mediated cytosine methylation and C/G-T/A base pair transitions can cause the degeneration of functional reading frames, as observed in many sequenced EPRV copies, e.g. in NsEPRV (Jakowitsch et al. 1999; Sec. 4.6.8). Large-scale elimination events have been reported for copies of *NtoEPRV* after the formation of synthetic allotetraploids (Skalická et al. 2005) and most likely also happen in nature (Matzke et al. 2004). Such instability is probably caused by chromosomal rearrangements and might be related to the recombinogenic effects of closely associated *Ty3-gypsy* elements (*Metaviridae*) (Gregor et al. 2004; Matzke et al. 2004; Skalická et al. 2005). Accumulation of deleterious mutations might then lead to the inactivation of EPRVs (Fig. 4.6), resulting in a transition from a horizontal to a vertical transmission mode for the EPRV.

Becoming part of the plant genome, the viral sequences can serve as a reservoir contributing to genetic variability of exogenous virus via recombination or (if still functionally intact) activation (Fig. 4.6). Comparable events might be partially responsible for the unexpected diversity of BSV-like isolates from Uganda (Harper et al. 2005).

4.6.7 Maintenance

The presence of EPRVs in such a wide range of plant species suggests that they confer some selective advantage. One such advantage could be protection from infection by related viruses by an RNA interference (RNAi) type of resistance (Hull et al. 2000; Mette et al. 2002; Geering et al. 2005b; Noreen et al 2007). Mette et al. (2002) observed the silencing and methylation of a transgene driven by an EPRV-derived promoter in tobacco, which harbours homologous promoter elements in its

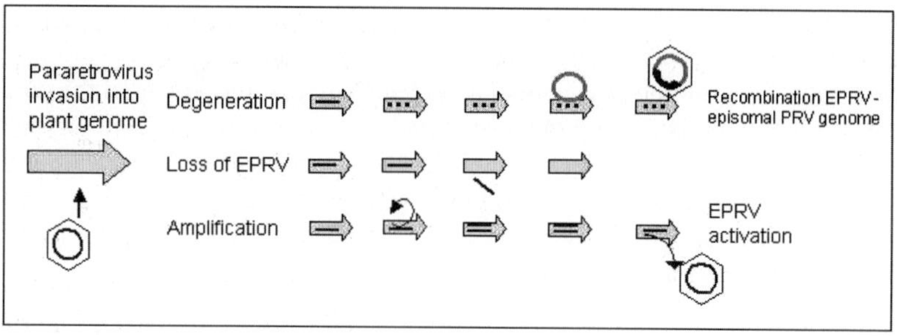

Fig. 4.6 Fate of the integrated pararetrovirus sequence. Genome invasion can be followed by the element's amplification, the degeneration of functional structures and/or large-scale elimination events. Conserved EPRV copies may still be activated, whereas degeneration leads to a strictly vertical transmission mode of the EPRV. Nevertheless, degenerate virus sequences could still contribute to genetic variability of exogenous viruses via recombination

genome. In contrast, the construct remained active and unmethylated in a genome lacking these EPRV sequences.

Since RNAi-mediated resistance does not require translation of the trigger, intact sequences capable of coding for functional and potentially harmful proteins would not be necessary. Consequently, there is little selection pressure for keeping beneficial EPRVs intact below a certain level of sequence identity. Consequently they would accumulate deletions, frameshifts, nonsense codons and other mutations (Jakowitsch et al. 1999; Kunii et al. 2004, Ndowora et al. 1999). Pathogenic EPRVs such as ePVCV, eTVCV and some different BSV integrants might represent an early stage in degradation of the sequence. However, in each of these three cases the EPRV can only be activated at a significant level in interspecific polyploids or natural interspecific *M. acuminate* × *M. balbisiana* hybrids, such as a wide range of plantains and cooking bananas, and therefore has been subject to natural selection to minimize or prevent activation in the original host.

It is possible for virus sequences to be subverted or incorporated for use by the host. For example, expression of *Polydnavirus* genes in Lepidoptera is essential for survival of the parasitoid's offspring (Schmidt et al. 2001) and retroviral and pseudogene insertion sites reveal the lineage of human salivary and pancreatic amylase genes from a single gene during primate evolution (Samuelsen et al. 1990). Likewise, virus sequences integrated into plant genomes could have significant beneficial effects on gene expression, for example, as promoters or as nonspecific translational transactivators.

4.6.8 Mobilization and Epigenetic Control

Some EPRVs can be spontaneously mobilized. It is very unlikely that such a mobilization occurs from single integrants, since these lack the terminal repeats required to produce a functional RNA as messenger and replicative intermediate, as for true retroviruses (Fig. 4.7a, b). However, mobilization is possible from tandemly arranged units, mimicking the situation of true retroviruses (Fig. 4.7c). Such a mechanism was suggested for ePVCV, where tandem arrangements have in fact been observed (Richert-Pöggeler et al. 2003). In the absence of such tandem arrangement, mobilization could occur by recombination from two or more incomplete but complementing transcripts during reverse transcription (Fig. 4.7d). Such a mechanism was suggested for endogenous BSV (Ndowora et al. 1999).

EPRVs are usually methylated, combined with repressive histones (deacetylated and H3K9 demethylated) and are consequently only negligibly transcribed (Noreen et al. 2007; Staginnus et al. 2007). Such transcripts are assumed to serve as precursors for small interfering RNAs (siRNAs) guiding TGS and post-transcriptional gene silencing (PTGS) (Baulcombe 2004; Almeida and Allshire 2005; Meins et al. 2005; Wassenegger 2005). Various observations support this hypothesis. Silencing and methylation of tobacco transgenes driven by an EPRV-derived promoter could

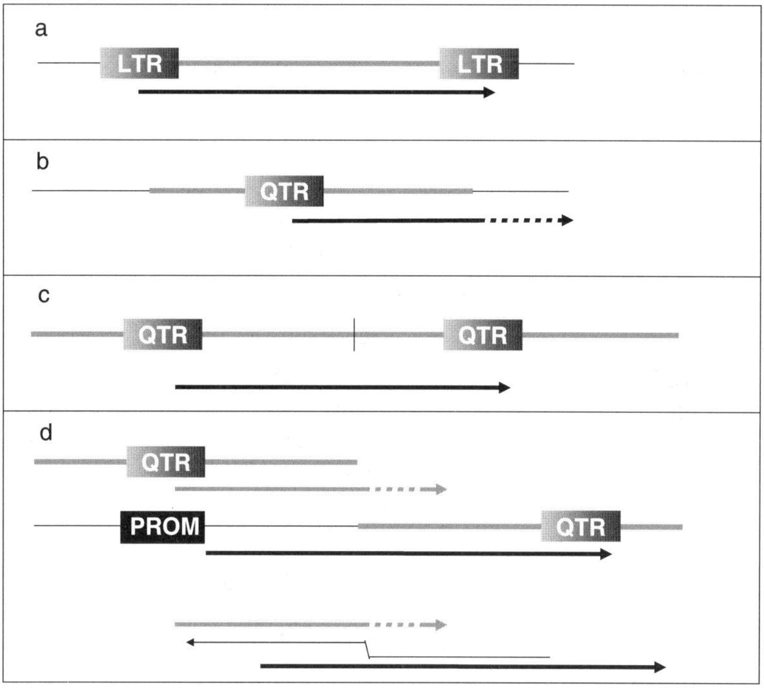

Fig. 4.7 Transcription of retrovirus and pararetrovirus genomes. **a** Mobilization of retroviruses by transcription of the integrated form from LTR to LTR leading to terminally redundant genomic RNA. **b** Incomplete transcription from randomly integrated pararetrovirus. **c** Transcription from tandemly repeated integrated pararetrovirus DNA leads to terminally redundant pararetrovirus RNA as for retroviruses. **d** Transcription of two separate integrants leading to pairs of transcripts that can recombine by template switching to form complete terminally redundant pararetrovirus RNA. *LTR* long terminal repeat containing promoter and polyadenylation signals, *QTR* quasi-LTR, solo pararetroviral sequences containing promoter (*PROM*) and polyadenylation sequences

be observed in the presence of homologous EPRVs in the genome (Mette et al. 2002). *Oryza* species carrying only a low copy number of eRTBV are much more susceptible to RTBV than species carrying a high copy number (Kunii et al. 2004). Findings from *P. hybrida* (Noreen et al. 2007) and *Lycopersicon esculentum* (now *Solanum lycopersicon*) species Staginnus et al. 2007) indicate the action of multiple dicerlike enzymes and point at multiple pathways. In fact, EPRV sequence methylation and histone H3K9 modification, typical for silenced chromatin, is found for ePVCV and LycEPRV (Noreen et al. 2007; Staginnus et al. 2007).

Although mobilization can occur spontaneously, there are various conditions that induce it more effectively. These include repetitive cuttings (Richert-Pöggeler et al. 2003), heat shock (Noreen et al. 2007), tissue culture (Dallot et al. 2001) and genetic hybridization (Lheureux et al. 2003). As noted earlier, ePVCV (Richert-Pöggeler et al. 2003), eTVCV (Lockhart et al. 2000) and some different BSV integrants (Harper et al. 1999; Lheureux et al. 2003) are usually not activated in the parent species but can be mobilized in hybrid species, perhaps by local relief of the silenced state, leading to systemic infections; the mobilized viruses are then not completely silenced by the endogenous copies. It is also possible that breeding processes such as interspecific hybridization or polyploidization suppress silencing mechanisms, leading to the expression of pathogenic EPRVs.

A recent model for epigenetic control of EPRVs (Staginnus and Richert-Pöggeler 2006) predicts that a host genome harbours methylated, heterochromatized (silent) copies and a few copies accessible to RNA polymerase that produce low levels of transcripts. These transcripts mostly lack the appropriate structure for the assembly of an infectious copy. Only in a few genomes, intact and functional structures of single copies are conserved. The transcripts (of intact or nonactivatable copies) provide templates required for siRNA production that induce or maintain epigenetic modifications like DNA methylation or heterochromatin formation of homologous loci via TGS and strictly control potentially infectious copies or invading exogenous virus. PTGS also is expected to be triggered (Fig. 4.8).

Genomic stress, such as hybrid formation or transient hypomethylation, weakens epigenetic control (Fischer et al. 2006). Accordingly the transcript levels increase and allow the production of terminally redundant RNA molecules that are crucial for viral replication, either by direct transcription of tandem arrays or via recombination steps. Systemic infection and symptoms of disease in the host plant result from and might be promoted by the production of a viral suppressor protein (Wang and Metzlaff 2005). Additionally, host factor(s) provided by parental genomes might cause expression of EPRVs in the hybrid (Lheureux et al. 2003).

4.6.9 Age

Transposon and viral integrants are essentially molecular fossils within the host genome. Detailed age analysis has been performed in the case of retrotransposons in maize (SanMiguel et al. 1998). Those are very abundant, exist as many different

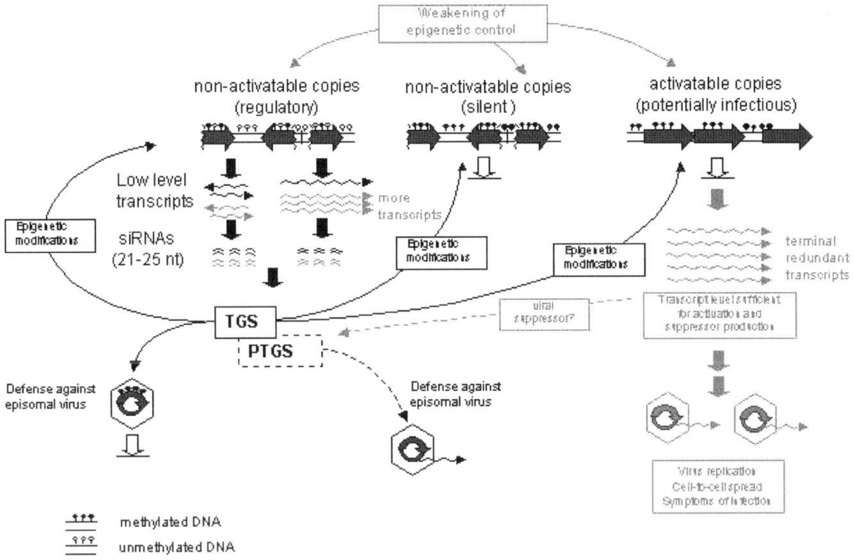

Fig. 4.8 Model for epigenetic control and a possible activation of EPRVs. *Black parts*: The genome mostly harbours methylated, heterochromatinized (silent) copies. A few copies are accessible to RNA polymerase and produce low levels of transcripts; however, these usually lack the appropriate structure for the assembly of an infectious copy. The transcripts provide templates required for small interfering (*siRNA*) production and induce or maintain epigenetic modifications (DNA methylation, heterochromatin formation) of homologous loci via transcriptional gene silencing (TGS), and strictly control potentially infectious copies or invading exogenous virus. In addition, post transcriptional gene silencing (PTGS) might be triggered (*dotted line*). *Red parts*: Epigenetic control is weakened, e.g. by hybrid formation or transient hypomethylation, which causes a general increase in transcript level. This allows the production of terminally redundant RNA molecules either directly from tandem arrays or via recombination steps which are necessary for viral replication. Additionally, the increased production of viral proteins might include a viral suppressor that weakens the epigenetic defence system. Both mechanisms lead to systemic infection and symptoms of disease in the host plant.

families, are incorporated as clusters and together make up to 50% of the maize genome. Each retrotransposon can be considered as a stratum that originated at a later time than the DNA flanking it. The key to dating the insertions is the LTR. It is possible to date these strata, because sequence divergence between the initially identical LTRs flanking a specific transposon should be proportional to the time that has been elapsed since its insertion. On the basis of these assumptions SanMiguel et al (1998) calculated that all the transposons around the *Adh1* gene locus have been inserted within the last six million years, most of them within the last three million years.

This type of analysis is less appropriate for EPRVs, since these are far less abundant than retrotransposons and do not have flanking LTRs. Thus, to determine EPRV age one has to access data on evolutionary differences in their hosts. Also, as mentioned before, there appears to be selection pressure that maintains the integrant and therefore it should coevolve with its host.

Some examples of how the age of integrants can be assessed are given below. Differences in EPRV presence and copy number within the genomes of tomato, tobacco, petunia, rice and banana demonstrate that invasion by pararetroviruses correlates with specific time points during speciation in the respective families (Kunii et al. 2004; Matzke et al. 2004; Geering et al. 2005b).

Kunii et al. (2004) noted that RTBV integrants must have existed in the rice AA genome before the differentiation of the *Japonica* and *Indica* subspecies, which occurred more than 8,000 years ago. However, since such integrants are restricted to the genus *Oryza* the integration event is likely to have occurred after that genus arose, about 130 million years ago.

There appears to be no commonality between BEV sequences of *M. accuminata* and *M. balbisiana*, suggesting that integration events took place after these species separated (Geering et al. 2005b), about one million years ago (Lescot et al. 2005). Some BEVs were found in all subspecies of *M. accuminata*, while others were from a restricted set of subspecies, indicating that integration was a continuing phenomenon during speciation. The greater diversity of BEVs in *M. balbisiana* could reflect a major radiation of badnaviruses in the region of origin of that *Musa* species (Jones 1999).

TVCV integrants are found in the genomes of *N. glutinosa*, *N. tabacum* and *N. rustica* from the Andean region of South America, but not in the genome of *N. clevelandii*, which is of North American origin, or those of *N. benthamiana* or *N. occidentalis*, which have Australian origin (Lockhart et al. 2000). Therefore, the integration event occurred after the dispersal of the progenitors of present-day *Nicotiana* species. It is suggested that the genus *Nicotiana* arose in what is now South America and that the separation into the three major subgenera, *Rustica* (South America), *Tabacum* (North America) and *Petunoides* (Australasia and Oceanic Pacific) took place after these areas had separated (Goodspeed 1947); the recent new classification of *Nicotiana* (Knapp et al. 2004) supports this geographic distribution. The separation of Gondwana which gave rise to South America and Australasia occurred about 50–60 million years ago.

As noted above, there are two distinct EPRV families in *N. tabacum*. These are derived from the progenitors of the allotetraploid *N. tabacum*, (NtoEPRV) and *N. sylvestris* (NsEPRV) and have less than 80% amino acid sequence identity; NtoEPRV is closely related to TVCV. As *N. tomentosiformis* and *N. sylvestris* are South American, a similar argument to that proposed for TVCV can be put forward for the age of NsEPRV and NtoEPRV. However, as noted by Matzke et al. (2004), native South Americans propagated various species by cuttings, a process that would have facilitated the maintenance of integrated sequences. Matzke et al (2004) also reported on the presence of sequences related to NtoEPRVs and NsEPRVs in other cultivated South American crops, including *Solanum tuberosum*, *Capsicum annum* and *Lycopersicon esculentum* (now *Solanum lycopersicon*). Thus, as for TVCV, it is likely that integration took place after the dispersal of these solanaceous species. The situation is the same for GRDs in *Nicotiana* species. As North America resulted from the breakup of Lauratia at around the same time as Gondwana separated, similar arguments apply to GRDs which are found in members of the *Tabacum* subgenus.

The overall presence of ePVCV found in *P. hybrida* cultivars analysed so far (Harper et al. 2003) indicates that genome invasion by the virus happened before *P. hybrida* was generated by artificial crosses almost 200 years ago (Sink 1984). Phylogenetic analysis of the genus *Petunia Jussieu*, that is endemic to South America, identified two major taxa, *Petunia sensu Wijsman* and *Calibrachoa*, which differ morphologically and in chromosome number (Ando et al. 2005) and diverged from other genera about 25 million years ago (Kulcheski et al. 2006). Integrated sequences were found in both the suggested parental crossing partners of *P. hybrida*, *P. axillaris* and *P. integrifolia*, (Richert-Pöggeler et al. 2003). More distantly related petunia species like *P. parodii* and *P. inflata* (Ando et al. 2005) do not contain any ePVCV sequences (Harper et al. 2002). The same is true for the one *Calibrachoa* species (*C. parviflora*) tested (Richert-Pöggeler, unpublished data). Thus, during speciation of petunia, multiple independent invasion events probably occurred in the recent past since the integrated sequences are still inducible in *P. hybrida* (Richert-Pöggeler et al. 2003; Noreen et al. 2007).

These estimates show that many EPRVs could be quite ancient. A further observation indicating the possibility of considerable antiquity was made by Hansen (2003), who found evidence for pararetrovirus-like sequences in the genomes of a number of dicotyledonous, monocotyledenous and gymnosperm genera.

4.7 Conclusions

There is clear evidence that plant DNA viruses can integrate into a host genome, and some of these integrations become fixed. These viruses are natural genetic engineers and may have been altering plant genomes since the origin of the plant kingdom. This is not a passive situation; the plant host has to cope with powerful regulatory sequences present within the viral genome and uses mechanisms similar to those controlling transposable elements that are parasites of any genome.

These integrated viral sequences can have unexpected consequences for crop improvement performed by breeding or genetic transformation. Host control of virus sequences can be affected in interspecific hybrids and may enable virus activation. Moreover, recombination between integrants and infecting viruses may create new viruses with modified biological properties such as host range or symptoms.

Overall, the impact of EPRVs on genome evolution and shape is becoming evident and indicates a close relationship between virus and host. Future research is necessary to provide more precise insights of such intimate symbioses and should be able to unravel possible functional roles of EPRVs in the regulatory pathways of the host.

Acknowledgements This article reflects a collaboration of the authors within the Paradigm project funded by EU framework V (QLRT 2001-02098). We thank B. Hohn, M. Matzke, F. Noreen, S. Kubis, J. Romero, N. Al-Kaff and P. Heslop-Harrison for fruitful discussions and their involvement in this project. T.H. thanks T. Boller, A. Wiemken and C. Koerner for laboratory space and the Friedrich Miescher Institute for office space.

References

Alberter B, Ali RM, Jeske H (2005) Replicative intermediates of Tomato leaf curl virus and its satellite DNAs. Virology 331:441–448

Almeida R, Allshire RC (2005) RNA silencing and genome regulation. Trends Cell Biol 15:251–258

Ando T, Kokubun H, Watanabe H, Tanaka N, Yukawa T, Hashimoto G, Marchesi E, Suarez E, Basualdo IL (2005) Phylogenetic analysis of Petunia sensu Jussieu (Solanaceae) using chloroplast DNA RFLP. Ann Bot (Lond) 96:289–297

Ashby MK, Warry A, Bejarano ER, Khashoggi A, Burrell M, Lichtenstein CP (1997) Analysis of multiple copies of geminiviral DNA in the genome of four closely related *Nicotiana* species suggest a unique integration event. Plant Mol Biol 35:313–321

Baulcombe D (2004) RNA silencing in plants. Nature 431:356–363

Bejarano ER, Khashoggi A, Witty M, Lichtenstein C (1996) Integration of multiple repeats of geminiviral DNA into the nuclear genome of tobacco during evolution. Proc Natl Acad Sci USA 93:759–764

Bennetzen JL (2005) Transposable elements, gene creation and genome rearrangement in flowering plants. Curr Opin Genet Dev 15:621–627

Berry C, Hannenhalli S, Leipzig J, Bushman FD (2006) Selection of target sites for mobile DNA integration in the human genome. PLoS Comput Biol 2:1450–1462

Bill CA, Summers J (2004) Genomic DNA double-strand breaks are targets for hepadnaviral DNA integration. Proc Natl Acad Sci USA 101:11135–11140

Blevins T, Rajeswaran R, Shivaprasad PV, Beknazariants D, Si-Ammour A, Park HS, Vazquez F, Robertson D, Meins F Jr, Hohn T, Pooggin MM (2006) Four plant Dicers mediate viral small RNA biogenesis and DNA virus induced silencing. Nucleic Acids Res 34:6233–6246

Botstein D (1980) A theory of modular evolution for bacteriophages. Ann N Y Acad Sci 354:484–491

Coffin JM, Hughes SH, Varmus HE (1997) Retroviruses. Cold Spring Harbor Press, Cold Spring Harbor

Dallot S, Acuna P, Rivera C, Ramirez P, Cote F, Lockhart BEL, Caruana ML (2001) Evidence that the proliferation stage of micropropagation procedure is determinant in the expression of *Banana streak virus* integrated into the genome of the FHIA 21 hybrid (*Musa* AAAB). Arch Virol 146:2179–2190

Day AC, Bergerano ER, Buck KW, Burrel M, Lichtenstein CP (1991) Expression of an antisense viral gene in tobacco confers resistance to the DNA virus Tomato golden mosaic virus. Proc Natl Acad Sci USA 88:6721–6727

Eickbush TH (1997) Telomerase and retrotransposons: Which came first? Science 277(5328):911–912

Fargette D, Konate G, Fauquet C, Muller E, Peterschmitt M, Tresh JM (2006) Molecular ecology and emergence of tropical plant viruses. Annu Rev Phytopathol 442:35–260

Fauquet CM, Mayo MA, Maniloff J, Desselberger U, Ball LA (2005) Virus Taxonomy. VIIIth report of the International Committee on Taxonomy of Viruses. Elsevier, San Diego

Ferber MJ, Montoya DP, Yu C, Aderca I, McGee A, Thorland EC, Nagorney DM, Gostout BS, Burgart LJ, Boix L, Bruix J, McMahon BJ, Cheung TH, Chung TK, Wong YF, Smith DI, Roberts LR (2003) Integrations of the hepatitis B virus (HBV) and human papillomavirus

(HPV) into the human telomerase reverse transcriptase (hTERT) gene in liver and cervical cancers. Oncogene 22:3813–3820
Fischer A, Hofmann I, Naumann K, Reuter G (2006) Heterochromatin proteins and the control of heterochromatic gene silencing in *Arabidopsis*. J Plant Physiol 163:358–368
Foster TM, Lough TJ, Emerson SJ, Lee RH, Bowman JL, Forster RL, Lucas WJ (2002) A surveillance system regulates selective entry of RNA into the shoot apex. Plant Cell 14:1497–1508
Froissart R, Roze D, Uzest M, Galibert L, Blanc S, Michalakis Y (2005) Recombination every day: Abundant recombination in a virus during a single multi-cellular host infection. PLoS Biol 3:e89
Geering ADW, Olszewski NE, Dahal G, Thomas JE, Lockhart BEL (2001) Analysis of the distribution and structure of integrated banana streak virus DNA in a range of *Musa* cultivars. Mol Plant Pathol 2:207–213
Geering AD, Pooggin MM, Olszewski NE, Lockhart BE, Thomas JE (2005a) Characterisation of Banana streak Mysore virus and evidence that its DNA is integrated in the B genome of cultivated Musa. Arch Virol 150:787–796
Geering AD, Olszewski NE, Harper G, Lockhart BE, Hull R, Thomas JE (2005b) Banana contains a diverse array of endogenous badnaviruses. J Gen Virol 86:511–520
Geijskes RJ, Braithwaite KS, Smith GR, Dale JL, Harding RM (2004) Sugarcane bacilliform virus encapsidates genome concatamers and does not appear to integrate into the *Saccharum officinarum* genome. Arch Virol 149:791–798
Gifford R, Tristem M. (2003) The evolution, distribution and diversity of endogenous retroviruses. Virus Genes 26:291–315
Goodspeed TH (1947) On the evolution of the genus Nicotiana. Proc Natl Acad Sci USA 33:158–171
Gregor W, Mette MF, Staginnus C, Matzke MA, Matzke AJ (2004) A distinct endogenous Pararetrovirus family in Nicotiana tomentosiformis, a diploid progenitor of polyploid tobacco. Plant Physiol 134:1191–1199
Gundersen-Rindal DE, Lynn DE (2003) Polydnavirus integration in lepidopteran host cells in vitro. J Insect Physiol 49:453–462
Gutierrez C (2000) Geminiviruses and the plant cell cycle. Plant Mol Biol 43:763–772
Hanin M, Paszkowski J (2003) Plant genome modification by homologous recombination. Curr Opin Plant Biol 6:157–162
Hansen CN (2003) Pararetrovirus-like sequences in the genomes of plants. PhD thesis, University of Leicester
Hansen CN, Heslop-Harrison JS (2004) Sequences and phylogenies of plant pararetroviruses, viruses and transposable elements. Adv Bot Res 41:165–193
Hansen CN, Harper G, Heslop-Harrison JS (2005) Characterisation of pararetrovirus-like sequences in the genome of potato (Solanum tuberosum). Cytogenet Genome Res 110:559–565
Harper G, Osuji JO, Heslop-Harrison JS, Hull R (1999) Integration of banana streak badnavirus into the Musa genome: molecular and cytogenetic evidence. Virology 255:207–213
Harper G, Hull R, Lockhart B, Olszewski N (2002) Viral sequences integrated into plant genomes. Annu Rev Phytopathol 40:119–136
Harper G, Richert-Pöggeler KR, Hohn T, Hull R (2003) Detection of petunia vein-clearing virus: model for the detection of DNA viruses in plants with homologous endogenous pararetrovirus sequences. J Virol Methods 107:177–184
Harper G, Hart D, Moult S, Hull R, Geering A, Thomas J (2005) The diversity of Banana streak virus isolates in Uganda. Arch Virol 150:2407–2420
Hegedüs K, Dallmann G, Balazs E (2004) The DNA form of a retroviroid-like element is involved in recombination events with itself and with the plant genome. Virology 325:277–286
Hohn T (1994) Recombination of plant pararetroviruses. In: J Paszkowski (ed) Homologous recombination and gene silencing in plants. Kluwer, Dordrecht, pp 25–38
Hohn T, Richert-Pöggeler KR (2006) Recent advances in DNA virus replication. In: Hetherton KR (ed) Recent advances in DNA virus replication. Research Signpost Transworld, Kerala

Hull R (1992) Genome organization of retroviruses and retroelements: evolutionary considerations and implications. Semin Virol 3:373–382

Hull R (2001) Matthews plant virology. Academic, San Diego

Hull R, Covey SN (1996) Retroelements: propagation and adaptation. Virus Genes 11:105–118

Hull R, Harper G, Lockhart B (2000) Viral sequences integrated into plant genomes. Trends Plant Sci 5:362–365

Jakowitsch J, Mette MF, van der WJ, Matzke MA, Matzke AJ (1999) Integrated pararetroviral sequences define a unique class of dispersed repetitive DNA in plants. Proc Natl Acad Sci USA 96:13241–13246

Jones DR (1999) Introduction to banana, abacá and enset. In: Jones DR (ed) Diseases of banana, abacá and enset. CABI, Wallingford, pp 1–36

Jones DT, Taylor WR, Thornton JM (1992) The rapid generation of mutation data matrices from protein sequences. CABIOS 8:275–282

Kalendar R, Tanskanen J, Immonen S, Nevo E, Schulman AH (2000) Genome evolution of wild barley (*Hordeum spontaneum*) by *BARE-1* retrotransposon dynamics in response to sharp microclimatic divergence. Proc Natl Acad Sci USA 97:6603–6607

Kapitonov VV, Jurka J (2001) Rolling-circle transposons in eukaryotes. Proc Natl Acad Sci USA 98:8714–8719

Kato M, Takashima K, Kakutani T (2004) Epigenetic control of CACTA transposon mobility in Arabidopsis thaliana. Genetics 168:961–969

Knapp S, Chase MW, Clarkson JJ (2004) Nomenclatural changes and a new sectional classification of *Nicotiana* (Solanaceae). Taxon 53:73–82

Kononov ME, Bassuner B, Gelvin SB (1997) Integration of T-DNA binary vector 'backbone' sequences into the tobacco genome: evidence for multiple complex patterns of integration. Plant J 11:945–957

Kroemer JA, Webb BA (2004) Polydnaviruses genes and genomes: emerging gene families and new insights into Polydnavirus replication. Annu Rev Entomol 49:431–456

Kulcheski FR, Muschner VC, Lorenz-Lemke AP, Stehmann JR, Bonatto SL, Salzano FM, Freitas LB (2006) Molecular phylogenetic analysis of *Petunia juss.* (Solanaceae). Genetica 126:3–14

Kumar A, Bennetzen JL (1999) Plant retrotransposons. Annu Rev Genet 33:479–532

Kunii M, Kanda M, Nagano H, Uyeda I, Kishima Y, Sano Y (2004) Reconstruction of putative DNA virus from endogenous rice tungro bacilliform virus-like sequences in the rice genome: implications for integration and evolution. BMC Genomics 5:80

Lafleur DA, Lockhart BEL, Olszweski NE (1996) Portions of Banana streak badnavirus genome are integrated in the genome of its host *Musa* sp. Phytopathology 86:100

Laten HM, Majumdar A., Gaucher EA (1998) *SIRE-1*, a *copia/Ty1*-like retroelement fromp soybean, encodes a retrovirial envelope-like protein. Proc Natl Acad Sci USA 95:6897–6902

Lescot M, Ciampi AY, Ruis M, Blanc G, Leebens-Mack J, Grasmeur O, D'Hunt A, da Silva FR, Ronning CM, Cheung F, Hans BJ, Alhoff R, Arbogast T, Hine E, Pappas G, Souza MT, Miller R, Glaszmann JC, Town CD, Piffanelli P (2005) Fresh insights into the Musa genome and its comparison with rice. http://pbil.univ-lyon1.fr/events/jobim2005/proceedings/P16Lescot.pdf

Lheureux F, Carreel F, Jenny C, Lockhart BE, Iskra-Caruana ML (2003) Identification of genetic markers linked to banana streak disease expression in inter-specific Musa hybrids. Theor Appl Genet 106:594–598

Lheureux F, Laboureau N, Muller E, Lockhart BEL, Iskra-Caruana ML (2007) Molecular characterization of *Banana streak virus acuminata vietnam* isolated from *Musa acuminata* siamea (banana cultivar). Arch Virol (in press)

Lockhart BE, Menke J, Dahal G, Olszewski NE (2000) Characterization and genomic analysis of tobacco vein clearing virus, a plant pararetrovirus that is transmitted vertically and related to sequences integrated in the host genome. J Gen Virol 81:1579–1585

Lovisolo O, Hull R, Rösler O (2003) Coevolution of viruses with hosts and vectors and possible paleontology. Adv Virus Res 62:325–379

Malik HS, Eickbush TH (1999) Modular evolution of the integrase domain in the Ty3/Gypsy class of LTR retrotransposons J Virol 73:5186–5190

Maori E, Tanne E, Sela I (2007) Reciprocal sequence exchange between non-retro viruses and hosts leading to the appearance of new host phenotypes. Virology 362:342–349

Matzke MA, Mette MF, Aufsatz W, Jakowitsch J, Matzke AJ (2000) Integrated pararetroviral sequences. Nat Biotechnol 18:579

Matzke M, Gregor W, Mette MF, Aufsatz W, Kanno T, Jakowitsch J, Matzke AJM (2004) Endogenous pararetroviruses of allotetraploid *Nicotiana tabacum* and its diploid progenitors, *N.sylvestris* and *N.tomentosiformis*. Biol J Linn Soc 82:627–638

Mayo MA, Jolly CA (1991) The 5′ nucleotide sequence of the genome of potato leaf roll virus RNA: evidence of recombination between virus and host RNA. J Gen Virol 72:2591–2595

Meins F, Si-Ammour A, Blevins T (2005) RNA silencing systems and their relevance to plant development. Annu Rev Cell Dev Biol 21:297–318

Mette MF, Kanno T, Aufsatz W, Jakowitsch J, van der Winden J, Matzke MA, Matzke AJ (2002) Endogenous viral sequences and their potential contribution to heritable virus resistance in plants. EMBO J 21:461–469

Mumm JP, Landy A, Gelles J (2006) Viewing single lambda site-specific recombination events from start to finish. EMBO J 25:4586–4595

Murad L, Bielawski JP, Matyasek R, Kovarik A, Nichols RA, Leitch AR, Lichtenstein CP (2004) The origin and evolution of geminivirus-related DNA sequences in *Nicotiana*. Heredity 92:352–358

Ndowora T, Dahal G, LaFleur D, Harper G, Hull R, Olszewski NE, Lockhart B (1999) Evidence that badnavirus infection in Musa can originate from integrated pararetroviral sequences. Virology 255:214–220

Noreen F, Akbergenov R, Hohn T, Richert-Pöggeler KR (2007) Distinct expression of endogenous *Petunia vein clearing virus* and the DNA transposon *dTph1* in two *Petunia hybrida* lines is correlated with differences in histone modification and siRNA production. Plant J (in press)

Ohba T, Yoshioka Y, Machida C, Machida Y (1995) DNA rearrangement associated with the integration of T-DNA in tobacco: an example for multiple duplications of DNA around the integration target. Plant J 7:157–164

Orend G, Linkwitz A, Doerfler W (1994) Selective sites of adenovirus (foreign) DNA integration into the hamster genome: changes in integration patterns. J Virol 68:187–194

Pahalawatta V, Droffel K, Pappo HR (2007) Seed transmission of *Daheia mosaic virus* in *Daheia punata* Plant Dis 91:88–91

Paszkowski J, Shillito RD, Saul M, Mandak V, Hohn T, Hohn B, Potrykus I (1984) Direct gene transfer to plants. EMBO J 3:2717–2722

Piffanelli P, Noa-Carrazana J. C, Lescot M, Benabdelmouna A, Dolezel J, Matsumoto T, Silva-Rosales L, Lheureux F, Teycheney P.-Y, Geering A, D'Hont A, Glaszmann J.-C, Sasaki T, Caruana M.-L (2005) Molecular analysis of *Banana streak virus* (BSV) "integrants" into the nuclear genome of *Musa balbisiana*. In: Abstracts of the XIIIth international congress of virology, San Francisco

Puchta H (2005) The repair of double-strand breaks in plants: mechanisms and consequences for genome evolution. J Exp Bot 56:1–14

Richert-Pöggeler KR, Shepherd RJ (1997) Petunia vein clearing virus: a plant pararetrovirus with the core sequence for an integrase function. Virology 236:137–146

Richert-Pöggeler KR, Noreen F, Schwarzacher T, Harper G, Hohn T (2003) Induction of infectious petunia vein clearing (pararetro) virus from endogenous provirus in petunia. EMBO J 22:4836–4845

Safár J, Noa-Carrazana JC, Virana J, Bartos J, Alkhimova O, Sabou X, Samikova H, Lhereux F, Caruana ML, Dolezei J, Piffanelli P (2004) Creation of a BAC resource to study the structure and evolution of the banana (*Musa balbisiana*) genome. Genome 47:1182–1191

Samuelson LC, Wiebauer K, Snow CM, Meisler MH (1990) Retroviral and pseudogene insertion sites reveal the lineage of human salivary and pancreatic amylase genes from a single gene during primate evolution. Mol Cell Biol. 10:2513–2520

SanMiguel P, Gaut BS, Tikhonov A, Nakajima Y, Bennetzen JL (1998) The paleontology of intergene retrotransposons of maize. Nat Genet 20:43–45

Schmidt O, Theopold U, Strand MR (2001) Innate immunity and evasion by insect parasitoids. BioEssays 23:344–351

Schwach F, Vaistij FE, Jones L, Baulcombe DC (2005) An RNA-dependent RNA polymerase prevents meristem invasion by Potato virus X and is required for the activity but not the production of a systemic silencing signal. Plant Physiol 138:1842–1852

Sink KC (1984) Taxonomy. In: Sink KC (ed) Petunia. Springer, New York

Skalická K, Lim KY, Matyasek R, Matzke M, Leitch AR, Kovarik A (2005) Preferential elimination of repeated DNA sequences from the paternal, *Nicotiana tomentosiformis* genome donor of a synthetic, allotetraploid tobacco. New Phytol 166:291–303

Staginnus C, Richert-Pöggeler KR (2006) Endogenous pararetroviruses: two-faced travelers in the plant genome. Trends Plant Sci 11:485–491

Staginnus C, Gregor W, Mette MF, Teo CH, Borroto-Fernandez EG, Laimer da Camara Machado M, Matzke MA, Schwarzacher T (2007) Endogenous pararetroviral sequences in tomato (Solanum lycopersicum) and related species. BMC Plant Biol 7:24

Steeves TA, Sussex IM (1989) Patterns in plant development. Cambridge University Press, Cambridge

Tanne E, Sela I (2005) Occurrence of a DNA sequence of a non-retro RNA virus in a host plant genome and its expression: evidence for recombination between viral and host RNAs. Virology 332:614–622

Tsuei DJ, Chang MH, Chen PJ, Hsu TY, Ni YH (2002) Characterization of integration patterns and flanking cellular sequences of hepatitis B virus in childhood hepatocellular carcinomas. J Med Virol 68: 513–521

Tuskan GA, Difazio S, Jansson S, Bohlmann J, Grigoriev I, Hellsten U, Putnam N, Ralph S, Rombauts S, Salamov A, Schein J, Sterck L, Aerts A, Bhalerao RR, Bhalerao RP, Blaudez D, Boerjan W, Brun A, Brunner A, Busov V, Campbell M, Carlson J, Chalot M, Chapman J, Chen GL, Cooper D, Coutinho PM, Couturier J, Covert S, Cronk Q, Cunningham R, Davis J, Degroeve S, Dejardin A, Depamphilis C, Detter J, Dirks B, Dubchak I, Duplessis S, Ehlting J, Ellis B, Gendler K, Goodstein D, Gribskov M, Grimwood J, Groover A, Gunter L, Hamberger B, Heinze B, Helariutta Y, Henrissat B, Holligan D, Holt R, Huang W, Islam-Faridi N, Jones S, Jones-Rhoades M, Jorgensen R, Joshi C, Kangasjarvi J, Karlsson J, Kelleher C, Kirkpatrick R, Kirst M, Kohler A, Kalluri U, Larimer F, Leebens-Mack J, Leple JC, Locascio P, Lou Y, Lucas S, Martin F, Montanini B, Napoli C, Nelson DR, Nelson C, Nieminen K, Nilsson O, Pereda V, Peter G, Philippe R, Pilate G, Poliakov A, Razumovskaya J, Richardson P, Rinaldi C, Ritland K, Rouze P, Ryaboy D, Schmutz J, Schrader J, Segerman B, Shin H, Siddiqui A, Sterky F, Terry A, Tsai CJ, Uberbacher E, Unneberg P, Vahala J, Wall K, Wessler S, Yang G, Yin T, Douglas C, Marra M, Sandberg G, Van de Peer Y, Rokhsar D (2006) The genome of black cottonwood, *Populus trichocarpa* (Torr & Gray). Science 313(5793):1596–1604

Vera A, Daros JA, Flores R, Hernandez C (2000) The DNA of a plant retroviroid-like element is fused to different sites in the genome of a plant pararetrovirus and shows multiple forms with sequence deletions. J Virol 74:10390–10400

Wang MB, Metzlaff M (2005) RNA silencing and antiviral defense in plants. Curr Opin Plant Biol 8:216–222

Wang Y, Tang X, Cheng Z, Mueller L, Giovannoni J, Tanksley SD (2006) Euchromatin and pericentromeric heterochromatin: comparative composition in the tomato genome. Genetics 172:2529–2540

Wassenegger M (2005) The role of the RNAi machinery in heterochromatin formation. Cell 122:13–16

Wentzensen N, Vinokurova S, von Knoebel Doeberitz M (2004) Systematic review of genomic integration sites of human papillomavirus genomes in epithelial dysplasia and invasive cancer of the female lower genital tract. Cancer Res 64:3878–3884

White KA, Nagy PD (2004) Advances in the molecular biology of Tombusviruses: gene expression, genome replication, and recombination. Prog Nucleic Acid Res Mol Biol 78:187–228

Wilson DM III, Thompson LH (2007) Molecular mechanisms of sister-chromatid exchange. Mutat Res 616:11–23

Xiong Y, Eickbush TH (1990) Origin and evolution of retroelements based upon their reverse transcriptase sequences. EMBO J 9:3353–3362

Yang W, Summers J (1999) Integration of hepadnavirus DNA in infected liver: evidence for a linear precursor. J Virol 73:9710–9717

Yano ST, Panbehi B, Das A, Laten H.M (2005) Diaspora, a large family of Ty3-gypsy retrotransposons in Glycine max, is an envelope-less member of an endogenous plant retrovirus lineage. BMC Evol Biol 5:30

Chapter 5
Viroids

Robert A. Owens

Abstract Viroids are the smallest known agents of infectious disease – small, highly structured, single-stranded, circular RNA molecules that lack detectable messenger RNA activity. Whereas viruses supply some or most of the genetic information required for their replication, viroids are regarded as "obligate parasites of the cell's transcriptional machinery" and infect only plants. Four of the nearly 30 species of viroids described to date contain hammerhead ribozymes, and phylogenetic analysis suggests that viroids may share a common origin with hepatitis delta virus and several other viroid-like satellite RNAs. Replication proceeds via a rolling-circle mechanism, and strand exchange can result in a variety of insertion/deletion events. The terminal domains of potato spindle tuber and related viroids, in particular, appear to have undergone repeated sequence exchange and/or rearrangement. Viroid populations often contain a complex mixture of sequence variants, and environmental stress (including transfer to different hosts) has been shown to result in a significant increase in sequence heterogeneity. The new field of synthetic biology offers exciting opportunities to determine the minimal size of a fully functional viroid genome. Much of the preliminary structural and functional information necessary is already available, but formidable obstacles still remain.

Robert A. Owens
Molecular Plant Pathology Laboratory, United States, Department of Agriculture/Agricultural Research Service, Beltsville, MD 20705
robert.a.owens@ars.usda.gov

M.J. Roossinck (ed.), *Plant Virus Evolution*.

5.1 Introduction

The first viroid disease to be studied by plant pathologists was potato spindle tuber. Nearly 50 years were to elapse between the discovery of its infectious nature and ability to spread in the field that led Schultz and Folsom (1923) to group potato spindle tuber disease with several other "degeneration diseases" of potatoes and the demonstration by Diener (1971) that the molecular properties of its causal agent, *Potato spindle tuber viroid* (PSTVd), differed fundamentally from those of conventional plant viruses. In the discussion section, this seminal paper focuses on the possible origin and evolution of viroids – as relatives of conventional viruses or the "missing link" between viruses and host genes. Initially, it was the small size of the viroid RNA genome (10^5 Da or less) and the absence of a protein capsid that appeared to set viroids apart from conventional viruses. Later, the circular structure and noncoding nature of viroids was seen as evidence for only a distant relationship between viroids and viruses. Shortly after, *Avocado sunblotch viroid* (ASBVd) was discovered to possess catalytic (i.e., ribozyme) activity. Diener (1989) proposed that viroids might be *living fossils* of a prebiotic RNA world where RNA molecules functioned as both genotype and phenotype.

Since the first complete viroid nucleotide sequence (that of PSTVd; Gross et al. 1978) was published nearly 30 years ago, much has been learned about the molecular biology of viroid replication and other aspects of viroid–host interaction. More than 1,100 complete sequences of PSTVd and other viroids are now available online from the Subviral RNA Database (http://subviral.med.uottawa.ca), and a number of studies have examined different aspects of viroid evolution. By examining the conceptual framework underlying existing studies of viroid evolution, this chapter attempts to identify fruitful areas for future studies. For information concerning other aspects of viroid molecular biology, interested readers should consult the monograph by Hadidi et al. (2003) and a series of recent reviews (Tabler and Tsagris 2004; Flores et al. 2005a; Ding et al. 2005; Daros et al. 2006; Ding and Itaya 2007).

5.2 Genome Structure and Replication Strategy

The Eighth Report of the International Committee on the Taxonomy of Viruses officially recognizes 29 viroid species and divides them into two families (the *Pospiviroidae* and the *Avsunviroidae*) containing a total of seven genera (Flores et al. 2005b; Table 5.1). All species in the family *Pospiviroidae* have a rod-like secondary structure that contains five structural/functional domains (Keese and Symons 1985) and replicate in the nucleus. Three of the four members of the *Avsunviroidae* have a branched secondary structure, and all replicate/accumulate in the chloroplast. All members of the *Avsunviroidae* contain hammerhead ribozymes in both the infectious (+) strand and complementary (–) strand RNAs. Figure 5.1 compares the secondary structures of PSTVd (rod-like, *Pospiviroidae*) and *Peach*

Table 5.1 Officially recognized viroid species (Eighth Report of the of the International Committee on the Taxonomy of Viruses)

Genus[a]	Species	Sigla	Reported variants[b]	Length (nt)
Family *Pospiviroidae*				
Pospiviroid	Potato spindle tuber	PSTVd	109	341–364
	Chrysanthemum stunt	CSVd	19	348–356
	Citrus exocortis	CEVd	86	366–475
	Columnea latent	CLVd	17	359–456
	Iresine	IrVd	3	370
	Mexican papita	MPVd	6	359–360
	Tomato apical stunt	TASVd	5	360–363
	Tomato chlorotic dwarf	TCDVd	2	360
	Tomato planta macho	TPMVd	2	360
Hostuviroid	Hop stunt	HSVd	144	294–303
Cocadviroid	Coconut cadang-cadang	CCCVd	8	246–301
	Coconut tinangaja	CCCVd	2	254
	Citrus bark cracking	CBCVd	6	284–286
	Hop latent	HLVd	10	255–256
Apscaviroid	Apple scar skin	ASSVd	8	329–333
	Apple dimple fruit	ADFVd	2	306
	Apple fruit crinkle	AFCVd[c]	29	368–372
	Australian grapevine	AGVd	1	369
	Citrus bent leaf	CBLVd	24	315–329
	Citrus dwarfing	CDVd	53	291–297
	Grapevine yellow speckle 1	GYSVd-1	49	365–368
	Grapevine yellow speckle 2	GYSVd-2	1	363
	Pear blister canker	PBCVd	18	314–316
Coleviroid	*Coleus blumei*-1	CbVd-1	9	248–251
	Coleus blumei-2	CbVd-2	2	295–301
	Coleus blumei-3	CbVd-3	3	361–364
Family *Avsunviroidae*				
Avsunviroid	Avocado sun blotch	ASBVd	83	239–251
Pelamoviroid	Chrysanthemum chlorotic mottle	CChMVd	21	397–401
	Peach latent mosaic	PLMVd	168	335–351
Elaviroid	Eggplant latent	ELVd	9	332–335

[a]Names of viroid genera are derived from those of the respective type species (listed first)
[b]Sequences available online from the Subviral RNA Database (http://subviral.med. uottawa.ca)
[c]Provisional species (not officially recognized)

latent mosaic viroid (PLMVd; branched, *Avsunviroidae*). With the possible exception of PLMVd, viroids do not appear to contain any modified nucleotides or unusual phosphodiester bonds.

Nucleic acid extracts from infected leaf tissue contain a variety of viroid-related RNAs of both polarities. Some of these molecules – especially those having a complementary or (–) strand polarity – are considerably longer than the infectious circular viroid (+) strand. Northern analysis using strand-specific probes and/or primer extension has shown that these molecules represent the intermediates expected for a "rolling-circle" mechanism of replication.

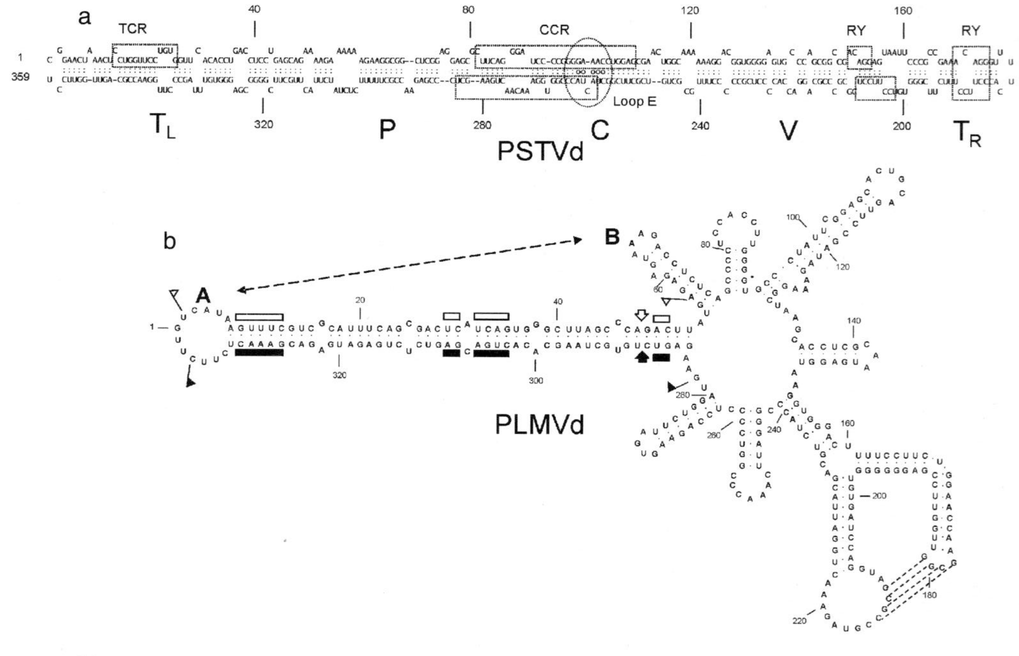

Fig. 5.1 **a** The rod-like secondary structure of *Potato spindle tuber viroid* (*PSTVd*; intermediate strain) showing the five domains characteristic of members of the family *Pospiviroidae:* the terminal left (T_L), pathogenicity (*P*), central (*C*), variable (*V*), and terminal right (T_R). The central conserved region (*CCR*) is located within the C domain and contains a UV-sensitive loop E motif with noncanonical base pairs (denoted by *circles*). The T_L domains of pospiviroids and apscaviroids contain a terminal conserved region (*TCR*), while those of hostuviroids and cocadviroids contain a terminal conserved hairpin (not shown). The T_R may also contain one or two copies of a protein-binding RY motif. **b** The branched secondary structure of *Peach latent mosaic viroid* (*PLMVd*: reference variant). (+) and (–) strand self-cleavage domains are indicated by *flags*, nucleotides conserved in most natural hammerhead structures by *bars*, and the self-cleavage sites by *arrows*. *Black symbols* and *white symbols* refer to plus and minus polarities, respectively. Nucleotides involved in a pseudoknot are indicated by *broken lines.* (Reproduced with permission from Malfitano et al. 2003)

Analysis of ASBVd-infected leaf tissue revealed the presence of monomeric circular RNAs of both polarities (Daros et al. 1994); thus, ASBVd (and presumably other avsunviroids) replicates via a symmetric rolling-circle mechanism (Fig. 5.2). Replication of PSTVd, in contrast, proceeds via an asymmetric rolling-circle mechanism in which progeny (+) strands are synthesized on a multimeric linear (–) strand template (Branch and Robertson 1984). The presence of hammerhead ribozymes in both strands of ASBVd allows its multimeric RNAs to cleave spontaneously, thereby releasing the corresponding linear monomers. Processing of longer-than-unit-length PSTVd (+) strand RNA requires that the central conserved region refold into a multihelix junction containing at least one GNRA tetraloop-hairpin followed by cleavage by an as-yet-unidentified host nuclease (Baumstark et al. 1997). A conserved loop E motif forms following the second cleavage reaction. Although evidence has been presented suggesting that monomeric linear PLMVd molecules can spontaneously circularize with the formation of a 2′,5′ phosphodiester linkage (Coté et al. 2001), circularization of most viroids appears to requires the action of a host RNA ligase.

A central question about viroid replication concerns the identity of the polymerase(s) involved. Inhibition of (+) and (–) strand PSTVd RNA synthesis by α-amanitin exhibits exactly the same dose–response effect in nuclear runoff experiments as host messenger RNA synthesis (Schindler and Muhlbach 1991), implicating host-DNA-dependent RNA polymerase II as the enzyme responsible for pospiviroid replication. Warrilow and Symons (1999) subsequently provided direct evidence for an association between RNA polymerase II and *Citrus exocortis viroid* (CEVd), showing that addition of a monoclonal antibody directed against the C-terminal domain of the largest subunit of RNA polymerase II results in immunoprecipitation of a nucleoprotein complex containing both (+) and (–) strand CEVd RNAs. Resistance of ASBVd RNA synthesis in permeabilized chloroplasts to

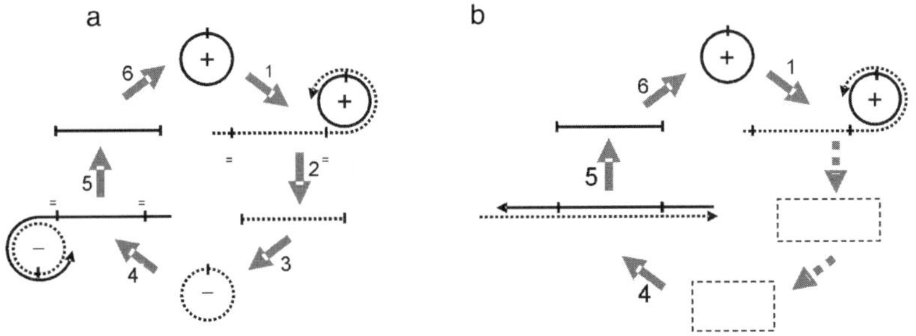

Fig. 5.2 Rolling-circle mechanism of viroid replication. **a** *Avocado sunblotch viroid* and other avsunviroids replicate via a *symmetric* mechanism in which cleavage of both multimeric (–) and (+) strand RNAs (steps 2 and 5) is mediated by ribozymes. Following cleavage, monomeric (-) strand RNA is circularized (step 3) before (+) strand synthesis begins. **b** PSTVd and other posiviroids replicate via an *asymmetric* mechanism in which multimeric (–) strand RNA synthesized by DNA-dependent RNA polymerase II (step 1) is not cleaved or ligated to prior to initiation of (+) strand synthesis (step 4)

tagetitoxin inhibition (Navarro et al. 2000) suggests that a nuclear-encoded RNA chloroplastic polymerase, and *not* the eubacteria-like RNA polymerase encoded by the plastid genome, is responsible for ASBVd strand elongation.

Although the existence of a defined cleavage site theoretically eliminates the need for a specific site for initiation of RNA synthesis, viroid replication appears to be promoter-driven. For ASBVd, both (+) and (–) strand synthesis initiate within AU-rich regions located in the terminal hairpin loops of the rod-like native structure. The nucleotide sequences around the ASBVd start sites bear a striking resemblance to the promoter sequences of certain chloroplast genes transcribed by the same nuclear-encoded RNA polymerase believed responsible for ASBVd replication (Navarro et al. 2000). Transcription of PSTVd (+) strands by RNA polymerase II starts at either position 359 or position 1 in the left terminal loop (Kolonko et al. 2006). Exactly how either of these viroids redirects the respective host-DNA-dependent RNA polymerase to accept its quasi-double-stranded RNA genome as a template is not known.

5.3 Evolutionary Relationships Among Subviral RNAs

Only two characters – an arbitrary level of less than 90% sequence similarity and differences in at least one biological property – are sufficient to distinguish the 29 officially recognized viroid species listed in Table 5.1. The broad outlines of this classification scheme first began to emerge in 1985, when Keese and Symons compared the sequences of eight viroids; i.e., one avsunviroid (ASBVd) and seven different pospiviroids. In the resulting model, PSTVd and other pospiviroids were proposed to contain five structural and functional domains. As shown in Fig. 5.1, these domains include (1) a central domain containing a conserved pair of inverted repeats and involved in viroid replication, (2) flanking pathogenicity and variable domains, and (3) two terminal domains that are interchangeable between viroids. The presence of partial sequence duplications in the right terminal domain of *Coconut cadang-cadang viroid* (CCCVd) was cited as evidence for the importance of RNA rearrangement/recombination in viroid evolution, suggesting that recombination with host RNAs may play an important role in the origin of viroids.

As nucleotide sequence information continued to accumulate, Elena and colleagues used phylogenetic analysis to examine two related topics, evolutionary relationships among viroids and the possible monophyletic origin of viroids and viroid-like satellite RNAs. Their initial study (Elena et al. 1991) included the viroid-like domain of hepatitis δ virus RNA and supported the previously proposed monophyletic origin for all subviral RNAs (Diener 1989). The resulting taxonomic classification contained both of the presently accepted viroid families as well as four of the five currently accepted *Pospiviroidae* genera (i.e., all except the coleviroids). When Jenkins et al. (2000) raised questions about the sequence alignment underlying this analysis, arguing that the sequence similarities needed to infer a reliable phylogeny were not present, Elena et al. (2001) reexamined these questions using improved techniques and a larger set of sequence data. Figure 5.3 shows the neighbor-joining tree produced by this reanalysis.

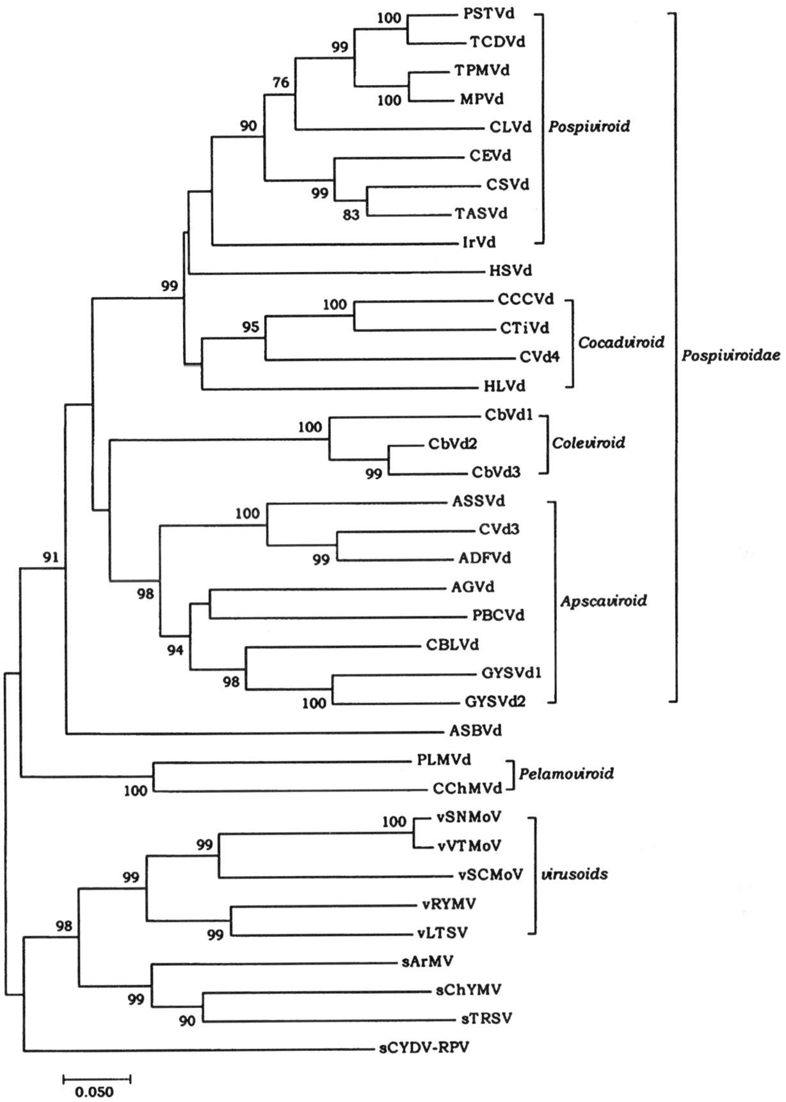

Fig. 5.3 Neighbor-joining phylogenetic tree obtained from an alignment of viroid and viroid-like RNA satellites. The distance matrix was obtained according to Hasegawa et al. (1985) but with gamma-distributed rates among sites (16 categories). Bootstrap values based on 1,000 random replicates, and only values greater than 70% are shown; less well supported branches can be collapsed. See Table 5.1 for viroid abbreviations. Viroid-like satellite RNAs: *vLTSV* (lucerne transient streak virus); *vRYMV* (rice yellow mottle virus); *vSCMoV* (subterranean clover mottle virus); *vSNMoV* (*Solanum nodiflorum* mottle virus); *vVTMoV* (velvet tobacco mottle virus); *sTRSV* (tobacco ringspot virus); *sArMV* (*Arabis* mosaic virus); *sChYMV* (chicory yellow mottle virus); *sCYDV-RPV* (cereal yellow dwarf virus-RPV). (Reproduced with permission from Elena et al. 2001)

The tree shown in Fig. 5.3 reproduces all groupings observed in the earlier analysis (Elena et al. 1991). Members of the family *Pospiviroidae* form three groups, one that includes the *Pospiviroid, Hostuviroid,* and *Cocadviroid* genera plus two other groups containing members of either the *Apscaviroid* or *Coleviroid* genera. For the autocatalytic subviral RNAs, viroids in the family *Avsunviroidae* can be seen to occupy a position intermediate between members of the *Pospiviroidae* and the viroid-like satellite RNAs. Additional support for these broad groupings was obtained by likelihood-mapping analysis, but key to the entire analysis was its two-stage alignment strategy. Sequences of viroids and viroid-like satellite RNAs were aligned separately using CLUSTAL-X and then manually edited to preserve local similarities; these partial alignments were then manually aligned before CLUSTAL-X was used to realign dissimilar regions and maximize overall similarity. An important anchor in this final alignment is a GAAA motif found in both the catalytic core of hammerhead ribozymes and the upper central conserved region of pospiviroids (Diener 1989).

5.4 Possible Roles of Conserved Sequence Motifs in Viroid Evolution

As shown in Fig. 5.1 and first described by Branch et al. (1985), the central domain of PSTVd (as well as other pospiviroids, apscaviroids, and possibly cocadviroids) contains a loop E motif similar to those found in a wide variety of other RNAs (Leontis and Westhof 1998). During rolling-circle replication, this loop E motif forms after cleavage of the nascent multimeric RNA and prior to monomer ligation. Loop E also contains a GAAA motif that is conserved among all hammerhead ribozymes. The terminal left domain of pospiviroids, apscaviroids, and coleviroids contains a terminal conserved region (Koltunow and Rezaian 1988), while those of hostuviroids and cocadviroids contain a terminal conserved hairpin (not shown). The terminal right domain of pospiviroids contains one or two copies of a protein-binding RY motif containing the sequence AGG/CUUUC (Gozmanova et al. 2003).

As first noted by Keese and Symons (1985), the terminal domains of pospiviroids appear to have been the target of repeated sequence exchange/rearrangement. For example, *Tomato apical stunt viroid* (TASVd) shares 73% overall sequence identity with CEVd, but the terminal right domains of these two viroids are only 46% similar. Although TASVd is less similar to PSTVd (only 64% overall similarity), their terminal right domains are highly (i.e., 90%) similar. Similar exchanges involving the terminal left domain of *Tomato planta macho viroid* (TPMVd) was also proposed. *Columnea latent viroid* (CLVd) appears to be a natural mosaic of viroid sequences that exhibits extensive sequence similarities to PSTVd and related viroids but has a central conserved region identical to that of *Hop stunt viroid*(HSVd; Hammond et al. 1989). These sorts of sequence rearrangements could be generated either by strand scission and ligation or, more likely, by discontinuous transcription.

Upon prolonged infection, both CCCVd (Haseloff et al. 1982) and CEVd (Szychowski et al. 2005) give rise to mixtures of larger RNA molecules containing duplications of part or all of the terminal right domain. In the case of CEVd, short AGCU tetrads provide possible sites for strand exchange by a "jumping polymerase," and the host plant plays a critical role in the origin and stability of specific duplications. Figure 5.4 shows the structures of CEVd-tomato and two of the longer CEVd-related RNAs, one containing an asymmetric duplication. Symmetrical sequence duplications like those found in CEVd D-87 create an additional binding site for Virp1, the viroid-binding host protein from tomato (Gozmanova et al. 2003). As shown in Fig. 5.1, wild-type (i.e., nonduplicated) PSTVd also contains two RY binding sites, and mutational analysis suggests that the ability of one or both of these sites to interact with Virp1 may be essential for some aspect of viroid transport (reviewed in Tabler and Tsagris 2004).

Spontaneous deletions have also been observed during viroid evolution in vivo. For example, constitutive expression of a noninfectious 350-nt PSTVd RNA containing a small deletion just upstream from the central conserved region in transgenic tobacco gave rise to a 341-nt PSTVd variant capable of independent replication in tomato (Wassenegger et al. 1994). The authors speculate that this rare event – observed in only a single plant expressing a (+) strand RNA transcript – was the result of either (1) host-DNA-dependent RNA polymerase being forced to jump over unpaired nucleotides in the quasi-double-stranded template or (2) a template repair process involving excision of unpaired nucleotides opposite the original deletion, followed by religation. Whether this was an all-or-none event is not known, but the net effect was to remove a large bulge loop in the opposing strand created by the original deletion in the PSTVd complementary DNA (cDNA).

As discussed by Schuster (2001), RNA molecules have many theoretical advantages over proteins for studies of the relationship between genotype and phenotype. These relationships are many-to-one and, thus, give ample room for neutrality. This is especially true for viroids where the rod-like (pospiviroids) or branched (avsunviroids) secondary structure predicted by computer calculation appears to be strongly conserved. Recently, Sanjuán et al. (2006) compared the "mutational robustness" of all 29 viroid species by calculating the structural effects of all possible single nucleotide point mutations. An evolutionary trend toward increased structural robustness was detected during viroid radiation, giving support to the presumed adaptive value of robustness. Genomic redundancy (see above) appears to contribute to structural stability, and the differences in robustness observed between the two viroid families can be explained by the relative fragility of the branched structure of members of the *Avsunviroidae*.

5.5 Structure of Viroid Quasispecies

Like many RNA viruses that infect plants or animals, individual viroids exist as complex populations of often closely related sequence variants in vivo. A number of studies have examined natural variability within viroid populations, and the

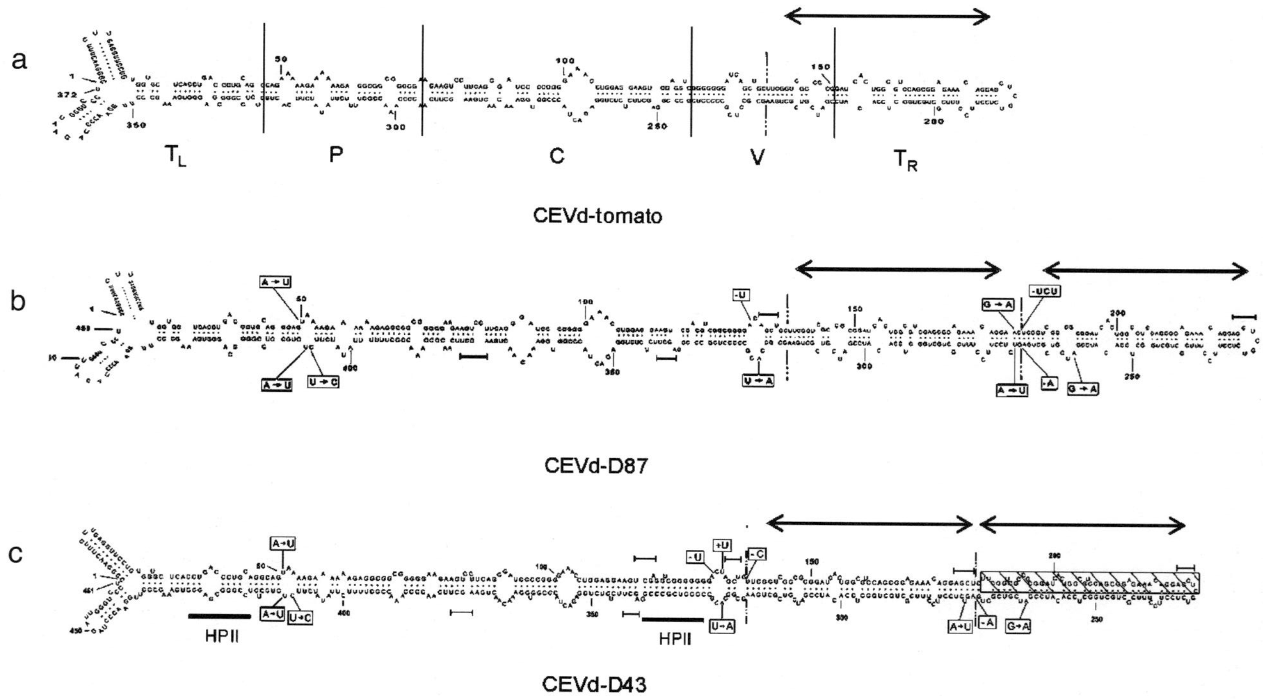

Fig. 5.4 Structures of selected CEVd sequence variants. **a** CEVd-tomato (the parental sequence); **b** CEVd D-87 containing a symmetric 87-nt duplication; and **c** CEVd D-43 containing an asymmetric 43-nt duplication. Nucleotides necessary to restore the symmetry of this asymmetric duplication are enclosed in a *cross-hatched box*. Positions of the five structural domains, terminal repeated regions (*horizontal arrows*), the AGCU tetrads (*horizontal bars*), and the stems of secondary hairpin II (*HPII*) are indicated. (Modified with permission from Fig. 2 in Szychowski et al. 2005)

Subviral RNA Database now contains the complete sequences of more than 1,100 viroid variants. In many cases, multiple sequence variants have been isolated from a single infected plant. Here, I focus on those viroids (e.g., HSVd, CEVd, PSTVd, and PLMVd) where (1) sequence information is abundant and (2) this information has been used to ask important questions about viroid–host interaction.

5.5.1 Origin of Viroid Diseases

As discussed by Diener (1996), several viroid diseases appear to be the result of chance transfer from an unknown (and possibly symptomless) wild host to large-scale monocultures of susceptible, genetically identical crop plants. The identity of these viroid reservoirs remains mysterious. Two groups of studies illustrate how the properties of different viroid variants have been used to explain the origin of specific viroid diseases.

The first example concerns hop stunt disease. Since its 1977 discovery as the causal agent of hop stunt disease, HSVd has been isolated from a wide variety of herbaceous and woody hosts (reviewed in Sano 2003). Two different groups have published phylogenetic analyses of known HSVd sequence variants (Kofalvi et al. 1997; Sano et al. 2001), and these can be divided into five clusters on the basis of the original host species:

1. A plum–peach–almond–apricot cluster
2. A German grapevine cluster
3. A general grapevine–hop cluster
4. A US citrus cachexia cluster
5. A general citrus–cucumber cluster

Figure 5.5 shows the neighbor-joining tree constructed by Sano et al. (2001), who used these relationships to propose a possible explanation for the origin of hop stunt disease in Japan.

Hops are native to the Mediterranean and Caucasus regions and were initially introduced to Japan from Europe and the USA at the end of the nineteenth century. Modern Japanese cultivars have been developed from these parental materials by crossing and selection. Hop stunt disease was first recognized in Nagano and Fukushima prefectures in the 1940s–1950s. So far, it has not been observed in either Europe or the USA. Thus, it would appear that the causal agent of hop stunt disease was introduced into hop from another host approximately 40–50 years after the introduction of hops into Japan. The close relationship between hop and grapevine isolates of HSVd shown in Fig. 5.5 strongly suggests that hop stunt disease originated when a sequence originally infecting grapevine invaded a new host (i.e., hop). Because phylogenetic analysis indicates that most HSVd-hop sequences form a single clade, the movement of HSVd-g from grapevine to hop must have been a unique (or at least an extremely rare) event. Hop gardens are often found adjacent to vineyards in Nagano and Fukushima prefectures, and farmers there seem to like

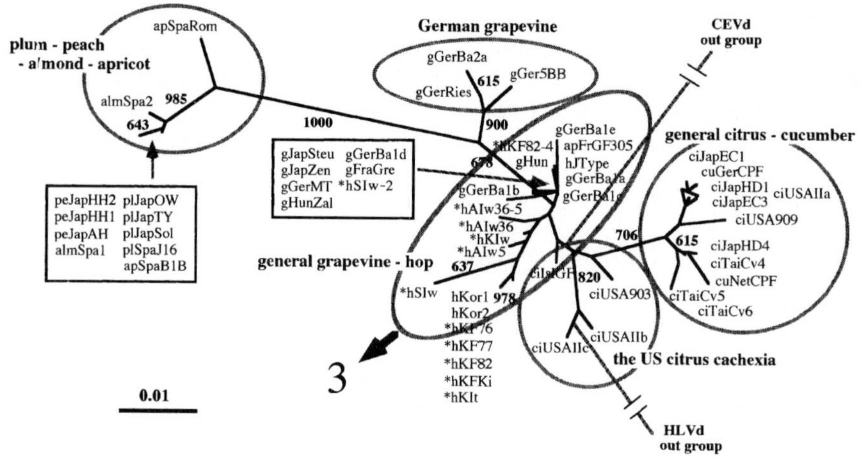

Fig. 5.5 Neighbor-joining analysis of HSVd variants recovered from a variety of hosts. CEVd and HLVd have been used as outgroups, and names of individual isolates are abbreviated. Lowercase letters indicate the original host plant: *h* hop, *g* grapevine, *ci* citrus, *pl* plum, *pe* peach, *al* almond, *ap* apricot, and *cu* cucumber; isolate names or numbers follow. Values on the nodes are confidence level estimates from bootstrap sampling (1,000 replicates). (Reprinted with permission from Sano et al. 2001 who sequenced the isolates denoted by *asterisks*)

planting a grapevine in a corner of their plowed field or hop garden. Once established, the viroid would be easily transmitted mechanically from hop to hop through contaminated tools because the titer in hops is more than 10 times higher than that in grapevine.

Another viroid that is 93–98% similar to AFCVd has been isolated from hops growing in Akita Prefecture, where apple cultivation is widespread. As described by Sano et al. (2004), phylogenetic analysis suggests that accumulation of host-specific sequence variation following isolation in different host species may be leading to the divergence of two viroid species from a common ancestor. A similar scenario has been proposed to explain several outbreaks of PSTVd in commercially grown tomatoes.

In 1990, Puchta et al. reported the isolation of a new strain of PSTVd from tomatoes being grown in two experimental greenhouses in the Netherlands. PSTVd-N contained only 356 nt and differed in sequence from the intermediate strain of PSTVd at 21 of 360 positions in a pairwise alignment (i.e., 5.8%). The same viroid was detected in healthy-looking pepino plants (*Solanum muricatum*) being grown in the same greenhouses; the seeds of these pepinos had been imported from New Zealand and Crete. Shortly thereafter, Behjatnia et al. (1996) reported the isolation of a very similar PSTVd variant from a wild *Solanum* sp. coinfected with two geminiviruses related to the Australian strain of *Tomato leaf curl virus* and growing in the Northern Territory of Australia. In recent years, infections of both field-grown (New Zealand) and greenhouse-grown (UK) tomatoes involving additional PSTVd variants related to PSTVd-N have been reported. Unlike the several host-specific groupings observed with HSVd (Fig. 5.5), unpublished phylogenetic analyses indicate that the PSTVd variant population is relatively unstructured. The one exception is PSTVd-N and the small group of related variants that appear to have originated in Oceania. It appears likely that one or more *Solanum* spp. act as a symptomless reservoir from which these variants periodically emerge to infect tomato (Verhoeven et al. 2004).

A final example involves the possible origin of potato spindle tuber disease. As described by Martínez-Soriano et al. (1996), the cultivated potato (*Solanum tuberosum* L.) originated in the Andes of South America, but attempts to isolate PSTVd or a similar viroid from other solanaceous species growing in this area, including some known to have been used in potato breeding, have been unsuccessful. Solanaceous plants native to Mexico have also been used in potato breeding, and one such species (i.e., *Solanum cardiophyllum* Lindl.) was shown to harbor *Mexican papita viroid* (MPVd), a previously unknown pospiviroid most closely related to TPMVd. These authors suggest that potatoes grown in the USA may have become infected by chance transfer of MPVd or a related viroid from endemically infected wild solaneous germplasm that was imported from Mexico in the late nineteenth century in efforts to identify sources of resistance to late blight. For this scenario to be correct, sequence evolution must have proceeded very rapidly after transfer to potato, however. While the two most distantly related variants in the PSTVd species differ by approximately 6%, PSTVd and MPVd exhibit only 80% sequence identity. The original host of PSTVd has yet to be identified.

5.5.2 Molecular Conformation and Disease Induction

Mild and severe strains of PSTVd (as well as several other viroids) differ only slightly in nucleotide sequence (Dickson et al. 1979), and much effort has been expended to determine how as few as one or two substitutions can have such dramatic biological effects. Less widely appreciated is the variation in sequence complexity exhibited by different viroid isolates. Pioneering studies by Visvader and Symons (1985) showed that both mild and severe isolates of CEVd contain a complex mixture of sequence variants, and characterization of three phenotypically different isolates of PSTVd yielded similar results (Góra et al. 1994). Additional studies have described the structure of ASBVd (Rakowski and Symons 1989; Schnell et al. 2001), CDVd (Owens et al. 2000), GYSVd (Rigden and Rezaian 1993; Polivka et al. 1996), *Hop latent viroid* (HLVd; Matousek et al. 2001) and HSVd (Palacio-Bielsa et al. 2004) populations. Arguably, the quasispecies concept has had the greatest influence on studies of PLMVd pathogenicity.

As discussed in Sec. 5.4, theoretical studies indicate that the branched secondary structure of PLMVd is more susceptible to disruption by point mutations than the rod-like conformation of members of the *Pospiviroidae* (Sanjuán et al. 2006). Naturally occurring isolates of PLMVd are divided into severe or latent strains depending on the appearance of leaf symptoms on seedlings of the peach indicator host GF-305 under greenhouse conditions. Extensive sequencing studies with PLMVd have been carried out by two different groups, and phylogenetic analysis of over 100 variants indicates that they can be divided into two groups sharing 85% sequence identity (Fig. 5.6). Group II variants include 99 variants showing 93% or more identity and can be further subdivided into three subgroups, each showing specific structural features. Variants in subgroup IIb appear to be the product of RNA recombination between members of subgroups IIA and IIC during coinfection (Hassen et al. 2007).

Taking advantage of this extensive sequence variation, both the Spanish and Canadian groups have used covariation analysis to refine their proposed secondary structure models for PLMVd. As shown in Fig. 5.1b, the right side of PLMVd contains a pseudoknot detected by nuclease probing in vitro (Bussière et al. 2000). Covariation analysis provides strong support for the existence of (1) several hairpin stems within this complex, highly branched region, and (2) a second pseudoknot that crosses the extended left side of the molecule containing the (+) strand (−) strand ribozymes (Ambros et al. 1998; Pelchat et al. 2000; Hassen et al. 2007). Induction of an extensive leaf chlorosis known as "peach calico" requires the presence of a 12–13-nt insertion in the hairpin loop containing positions 337 and 1. Although direct evidence is lacking, the presence of this insertion does not appear to interfere with pseudoknot formation; furthermore, the insertion spontaneously reappears upon prolonged passage of PLMVd molecules from which it had been deleted (Malfitano et al. 2003).

5.5.3 Stability of Natural Viroid Populations

Viroid isolates from woody perennials such as grapevine (Polivka et al. 1996) or citrus (Owens et al. 2000) may contain complex mixtures of sequence variants, likely reflecting the extended opportunity for sequence drift, RNA recombination, and/or reinfection. These naturally occurring viroid isolates (i.e., those not derived from cloned cDNAs) are usually quite stable when passaged in a single host under controlled conditions. On at least one occasion, however, a severe (and more rapidly replicating) sequence variant of PSTVd was observed to sweep through the quasispecies surrounding the intermediate green strain under house conditions (Gruner et al. 1995). Although the spontaneous conversion of PSTVd-Int to RG1 requires only three substitutions, mutational analysis suggests that PSTVd-Int occupies a comparatively steep peak within the fitness landscape (Owens et al. 2003).

In the case of CEVd, sequential passage of an isolate from a single sweet orange source through a series of alternate hosts (Etrog citron, *Gynura aurantiaca*, a *Lycopersicon esculentum* × *L. peruvianum* hybrid tomato, and disorganized hybrid tomato callus) resulted in a series of isolates differing in symptom expression, titer, and electrophoretic mobility (Semancik et al. 1993). Sequence analysis suggested the existence of a "tomato signature," a pattern of sequence changes shared by all isolates derived from hybrid tomato tissues. Two of these changes are located in the loop E motif of CEVd.

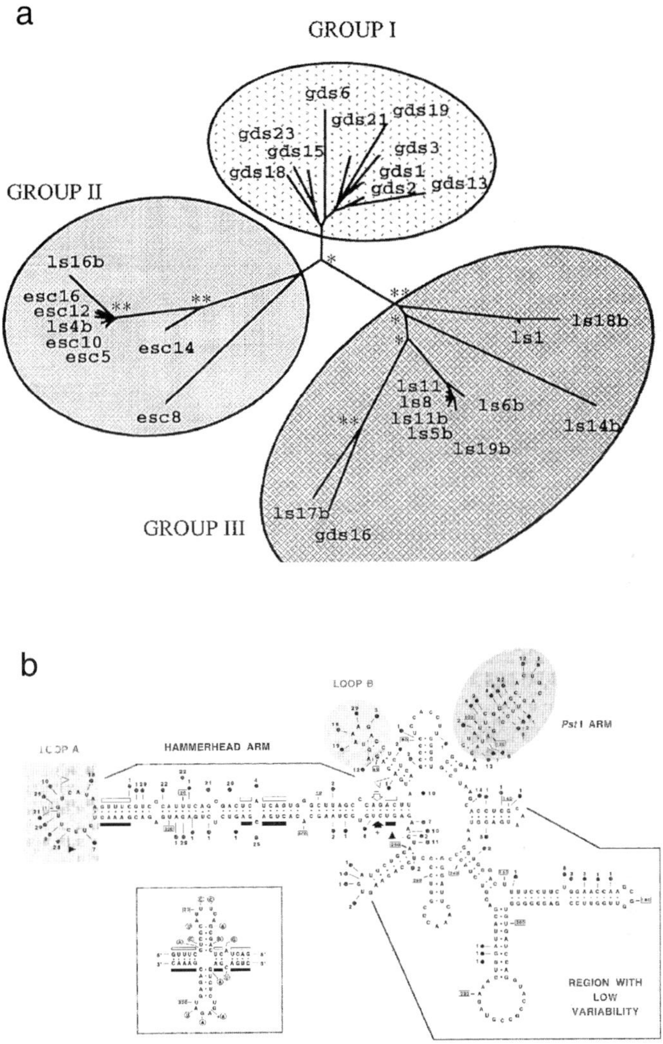

Fig. 5.6 Relationships among PLMVd sequence variants. **a** Dendrogram based on the genetic distances calculated between the 29 PLMVd sequence variants. Three groups are delineated, and *asterisks near nodes* indicate their statistical significance as determined by bootstrap analysis (100 replicates). *Double asterisks* node detected in 90–100% of replicates; *single asterisks* node detected in more than 50% of replicates. **b** Distribution of polymorphic positions along the secondary structure of PLMVd. Sequence changes (substitutions, insertions, or deletions) are marked by *circles*, and the number of variants in which each specific position is affected is indicated. Regions involved in forming (+) and (–) strand hammerhead structures are flanked by *flags*, the conserved nucleotides present in most natural hammerhead structures are indicated by *bars*, and the self-cleavage sites are indicated by *arrows*. *Black symbols* and *white symbols* refer to (+) and (–) polarities, respectively. The reference sequence is marked every 20 nt with *boxed numbers*. *Inset*: Alternative cruciform conformation for the hammerhead arm in the most stable secondary structure. (Reproduced with permission from Ambros et al. (1998))

Transfer of PSTVd from tomato (its most common experimental host) to tobacco (*Nicotiana tabacum* L.) results in the appearance of new sequence variants. Some of these variants contain a C/U substitution at position 259 (Wassenegger et al. 1996); another contained a U/A substitution at position 257 (Zhu et al. 2002). Both of these changes are located within the conserved core of loop E motif in PSTVd. A subsequent study by Zhong et al. (2006) considers the effects of these and other mutations in loop E on PSTVd replication and transport in light of the need to maintain its overall conformation. Because loop E motifs in cellular RNAs are well known to serve as important protein binding sites, it is easy to imagine how sequence changes in this portion of PSTVd or CEVd could result in dramatic shifts in quasispecies composition.

Thermotherapy is widely used to produce virus-free plant material (Mink et al. 1998), but the underlying mechanisms are not understood. Possibilities include (1) inactivation of intact virus particles and/or (2) an inhibition of virus replication/transport that allows the actively growing shoot apex to "outgrow" the much-reduced rate of virus spread. Viroid populations are also sensitive to heat stress. Using a combination of temperature gradient gel electrophoresis and DNA heteroduplex analysis to examine the effect of thermal stress on viroid populations in HLVd-infected hops (Matoušek et al. 2001) and PSTVd-infected *N. benethamiana* (Matoušek et al. 2004), Matoušek et al. showed that heat treatment is followed by a significant increase in sequence polymorphism (Fig. 5.7). Many variants contained multiple mutations, suggesting an accumulation of mutations during successive replication cycles; furthermore, approximately two thirds of all sequence changes in HLVd or PSTVd were located in the left side of secondary structure.

All HLVd variants tested were infectious, and all gave rise to complex progeny populations in hop. Most interesting were the populations of HLVd or PSTVd "thermomutants" observed after transfer to alternative hosts. Although no evidence of systemic infection could be detected when *Arabidopsis thaliana* was biolistically inoculated with PSTVd-Int strain, biolistic inoculation with a population of PSTVd thermomutants passaged through the intermediate host *Raphanus sativus* was followed by progeny accumulation to levels approximately 0.3% of those observed in tomato (i.e., detectable only by reverse-transcription PCR). Many of these molecules contained sequence changes in the upper portion of the central conserved region (a region known to play a key role in replication/cleavage), but these changes were not stably maintained when the individual variants were returned to tomato. Similar experiments involving transfer of HLVd thermomutants to tomato or *N. benthamiana* resulted in progeny accumulation to levels detectable by molecular hybridization (Matoušek 2003). Clearly, much remains to be learned about how the genetic diversity generated by thermal (and other possibly other environmental) stress may contribute to changes in viroid host range and speciation.

At high mutation rates, the fittest organisms are not necessarily those that replicate most quickly but rather those that show the greatest robustness against the generally deleterious effects of mutations. This phenomenon is sometimes termed "survival of the flattest." A recently published study by Codoñer et al. (2006) compared the ability of two viroids to compete against one another in coinfected chrysanthemum plants. One of the viroids tested (i.e., *Chrysanthemum stunt viroid*,

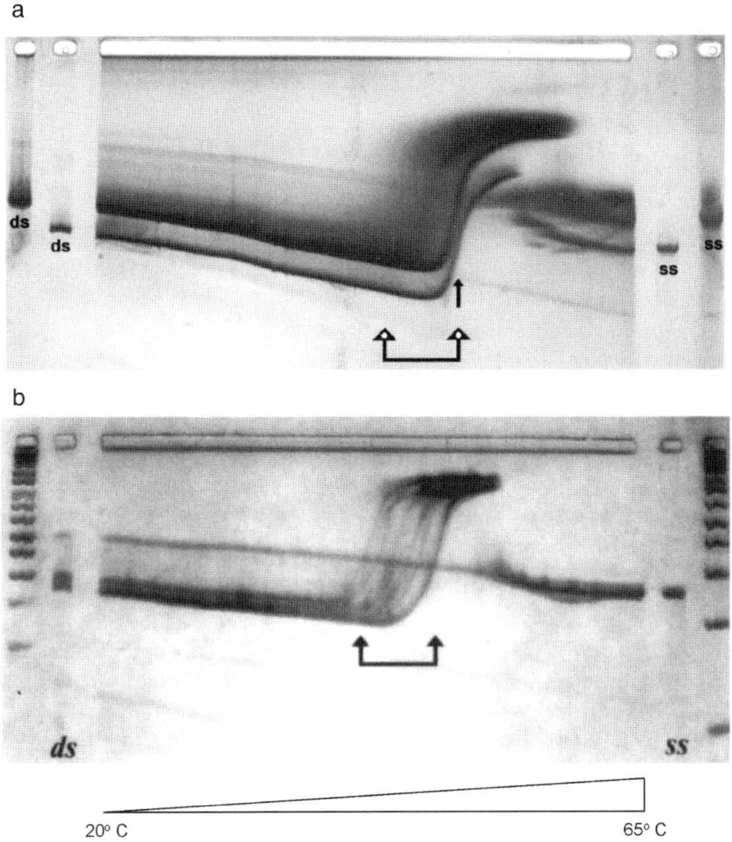

Fig. 5.7 Heat-induced accumulation of HLVd sequence variants. **a** Temperature gradient gel electrophoresis (TGGE) analysis of double-stranded HLVd complementary DNAs (cDNAs) prepared from viroid recovered from control and heat-treated hops. The control sample was loaded first; after 15 min of electrophoresis at 10°C, the sample from heat-treated plants was applied. The melting temperature of the major HLVd variant is indicated by the *single arrow*; additional cDNA species with melting temperatures in the range 42–48°C are indicated by the *double arrows*. These variants are present only in HLVd cDNA from heat-treated hop. **b** TGGE analysis of an HLVd population following heat treatment and three passages in *Nicotiana benthamiana*. HLVd cDNA was synthesized by reverse-transcription PCR, denatured, and allowed to form heteroduplexes before analysis. The range of melting temperatures of HLVd heteroduplexes is indicated by the *arrows*. A temperature gradient of 20–65°C was used for both analyses, and nucleic acids were visualized by silver staining. Double- and single-stranded portions of each gel pattern are indicated by *ds* and *ss*, respectively. (Modified with permission from Matousek et al. 2001 and Matousek 2003)

CSVd) replicates rapidly in the nucleus, and the progeny are genetically homogeneous. *Chrysanthemum chlorotic mottle viroid* (CChMVd), in contrast, replicates at a slower rate in the chloroplast, and progeny populations are highly variable. Under optimal growth conditions, CSVd outcompetes CChMVd as predicted by the Darwinian "survival of the fittest" paradigm. When the mutation rate is increased by exposure to UV radiation, however, the slowly growing (but more robust)

CChMVd predominates. These experimental results are consistent with computer simulations of competing viroid populations.

5.5.4 Generation of Populations From Individual Viroid Variants

Several different approaches have been used to monitor the genetic stability of individual viroid sequence variants in vivo. These include (1) inoculation with recombinant plasmid DNAs (Góra-Sochacka et al. 1997; Ambrós et al. 1999), (2) *Agrobacterium*-mediated introduction of nondisarmed recombinant Ti plasmids (Hammond 1994), and (3) *Agrobacterium*-mediated plant transformation (Wassenegger et al. 1994). When working with highly debilitated variants that are only weakly infectious, constitutive expression from an integrated transgene provides an effective means to detect the rare events that can restore viroid infectivity.

Góra et al. (1994) used a reverse-transcription PCR strategy to generate, in a single step, infectious full-length cDNAs from three phenotypically dissimilar isolates of PSTVd. When this method was applied to a "mild" isolate, only a single sequence variant was recovered. "Intermediate" and "severe" isolates yielded three and four variants, respectively. Not all of the variants recovered from the severe isolate produced severe symptoms when inoculated onto Rutgers tomato; thus, the presence of milder variants in a mixed inoculum may be masked by variants that are more severe. Follow-up studies (Góra-Sochacka et al. 1997) revealed that many of these naturally occurring PSTVd sequence variants were unstable when inoculated alone – sometimes disappearing within a single 5–6-week passage in tomato. This finding supports one of the basic tenets of the quasispecies theory, that mixtures of variants can complement each other, and hence the whole population is in essence a single entity analogous to an individual with thousands of alleles rather than just two. In most cases, the new variants detected induced symptoms that were less severe than those of the parent. The number of sequence changes detected in both studies was relatively limited, confined almost exclusively to the pathogenicity and variable domains (Fig. 5.1) with only a few changes located in the terminal right domain.

A similar study carried out by Ambrós et al. (1999) with PLMVd yielded very different results. As shown in Fig. 5.6, sequences comprising the PLMVd species can be divided into three groups. When GF-305 peach seedlings were slash-inoculated with a series of four cDNAs derived from PLMVd variants of differing pathogenicities, the resulting progeny contained a total of 33 sequence variants. The structure of the viroid populations derived from single PLMVd sequences differed according to the observed phenotype, and mutation frequencies were considerably higher than those reported for PSTVd (see above). While Malfitano et al. (2003) have shown that spontaneous mutational changes in PLMVd can include the appearance of a 12–13-nt insertion in the loop closing the hammerhead arm (Fig. 5.1b), the majority of changes observed were single nucleotide substitutions.

As discussed by the authors, this rapid evolution of PLMVd may be due to the involvement of a different (more error-prone?) RNA polymerase in its replication that contributes to the fluctuating symptoms observed with naturally occurring severe isolates of PLMVd. Polymorphisms appear to be limited by the need to maintain ribozyme activity and certain other structural and functional elements of PLMVd.

One important advantage of screening assays that involve mechanical inoculation of full-length viroid cDNAs or RNA transcripts is that the results are usually available within a few weeks. Many point mutations in PSTVd and other viroids, however, appear to abolish infectivity via mechanical inoculation. In some cases, these mutations have been shown to inhibit replication; in other cases, cell-to-cell or long-distance transport is disrupted (Qi et al. 2004). Placement of appropriately designed PSTVd cDNAs under the control of a constitutive or tissue-specific promoter followed by introduction of the resulting transgene into the nuclear genome of the host has made it possible to study the process by which severely debilitated variants regain systemic infectivity. No comparable experiments involving *Avsunviroidae* mutants and plastid transformation have been reported.

PSTVd mutants with changes in the terminal loops are not infectious when tomato cotyledons are mechanically inoculated with either cDNA or RNA transcripts (Hammond and Owens 1987). Nevertheless, when *Agrobacterium*-mediated inoculation was used to introduce a PSTVd mutant containing several changes in the right terminal loop into stem tissue, both replicative intermediates and viroid progeny could be recovered from gall and root tissues (Hammond 1994). Progeny were only occasionally detected in newly developing leaves, and sequence analyses revealed that the initial mutations in PSTVd-R were not stably maintained. Mutations in the right terminal loop do not abolish the ability of the PSTVd to replicate; rather, they appear to alter interaction with specific host components (e.g., VirP1; Gozmanova et al. 2003) and restrict the normal pattern of intercellular transport. Their phenotype resembles that of mutations elsewhere in the molecule that regulate transport across the bundle sheath–mesophyll boundary (Qi et al. 2004).

In a further refinement of this transgene-based approach, Wassenegger et al. (1994) used transformed tobacco to study the evolution of a second noninfectious PSTVd mutant *in planta*. The mutant contained a 9-nt deletion located immediately adjacent to the upper portion of the central conserved region in PSTVd-Int, and 20 independent primary transformants were screened by northern hybridization. Only one plant proved to be PSTVd-infected, and sequence analyses of the progeny revealed a surprising result. Instead of the 350-nt PSTVd variant encoded by the transgene, the infectious replicon contained only 341 nt. In addition to a complementary 9-nt deletion that restored the integrity of the rod-like native structure, the progeny also contained five additional point mutations. One of these point mutations was the same C259/U change in the loop E motif that is found in tobacco mechanically inoculated with PSTVd isolate KF440-2 (Wassenegger et al. 1996). Because the defective PSTVd RNA is continually produced in every cell nucleus and is available for RNA–RNA replication catalyzed by RNA polymerase II, it can be continuously mutated with selection acting at each step in the replicative pathway.

Unfortunately, the stochastic nature of this "repair" process makes it difficult to capture in its entirety.

5.6 Origin and Evolution of Viroids

Several possible origins for viroids have been proposed. Viroids could be primitive ancestors or highly degenerate derivatives of conventional viruses, but as discussed by Diener (1989), their unusual molecular structure and biological properties together with a lack of sequence similarity. Evolution argue against this possibility, of viroids from transposable elements, plasmids, or introns has also been proposed. At the present time, the balance of evidence suggests that viroids could represent "relics of precellular RNA evolution", and several reviews exploring this area have been published (Diener 1996, 2001, 2003).

In essence, the argument for viroid origin in the RNA world is straightforward: RNA is the only known biological macromolecule that can function as both genotype and phenotype, allowing evolution to occur in the absence of DNA or protein. As described by Diener (1989), a simple hammerhead ribozyme similar to those found in ASBVd and other members of the *Avsunviroidae* is theoretically capable of performing all the polymerization, cleavage, and ligation steps required for viroid replication. The circular structure of the viroid genome and the rolling-circle mechanism of replication eliminate the need for replication to initiate at a specific site; likewise, the apparently polyploid nature of viroid genomes (Juhasz et al. 1988) would have favored their survival under the error-prone conditions of the prebiotic world.

Figure 5.8 compares the structure of the first intermediate in the PSTVd cleavage-ligation pathway (Baumstark et al. 1997) with those of the hammerhead and hairpin ribozymes.The upper portion of the pospiviroid central conserved region contains a short sequence motif (GAAA) that is also present in hammerhead ribozymes (Diener 1989). Moving from the level of RNA primary/secondary structure to tertiary structure, however, one can see that pospiviroids share an even greater degree of similarity with ribozymes. The hairpin ribozyme found in (−) strand satellite RNA of *Tobacco ringspot virus* contains two domains that interact in the transition state. Like the central conserved region of pospiviroids, the loop B domain of the hairpin ribozyme also contains a loop E motif. Loop E motifs are found in many different contexts, often acting as "organizers" for multihelix loops in ribosomal RNAs (Leontis and Westhof 1998; Hendrix et al. 2005); in the case of the hairpin ribozyme, a conformational change in the loop E motif accompanies domain docking and is essential for catalysis (Hampel and Burke 2001). In addition to sequence-specific cleavage, the hairpin ribozyme also catalyzes RNA ligation. Recent experimental work with the hammerhead and hairpin ribozymes suggests that they have are more similar than previously thought (Burke 2002), and the possibility that viroids are "relics of precellular evolution" continues to be very much alive.

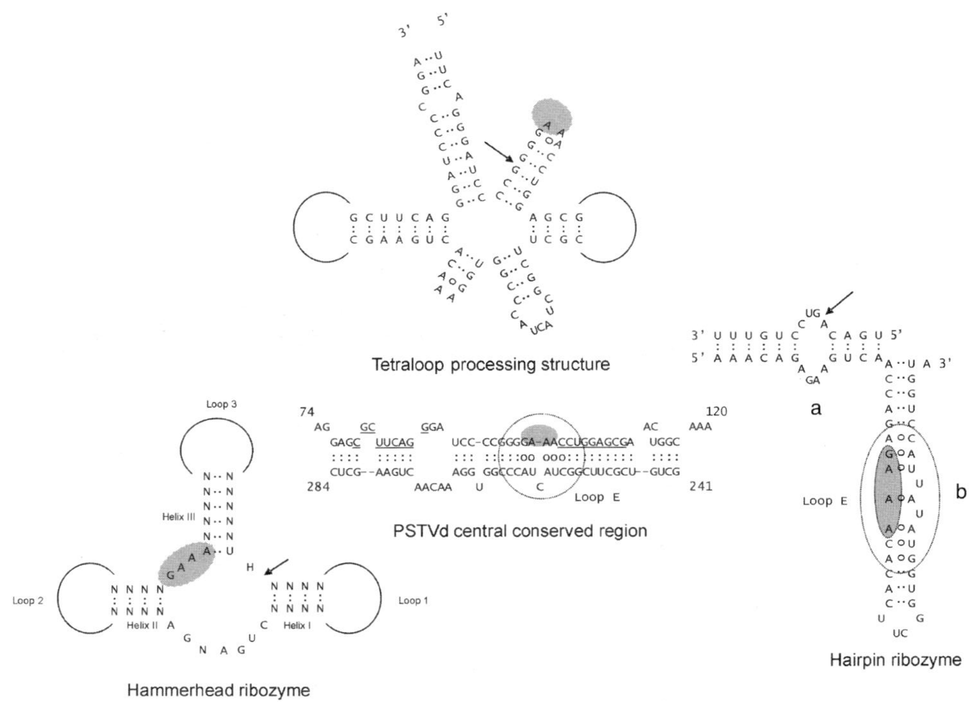

Fig. 5.8 Possible evolutionary relationships between viroids and ribozymes. The central conserved region of PSTVd and other *Pospiviroidae* is defined by a pair of inverted repeats (*underlined*) and contains a loop E motif (*dashed circle*). Loop E is not an internal loop but rather a highly structured array of non-Watson–Crick base pairs (denoted by *open circles*). Pospiviroids and hammerhead ribozymes (*lower left*) both contain a short conserved GAAA (shaded) sequence, suggesting to Diener (1989) the possible existence of a common ancestor in the prebiotic RNA world. The presence of a loop E motif in the hairpin ribozyme (*lower right*) provides additional support for such a relationship. The processing structure involved in the initial PSTVd cleavage (*above*) does not contain a loop E motif, however. Cleavage sites in the respective RNAs are denoted by *arrows*

5.7 Concluding Remarks

The now-routine determination of the nucleotide sequences of entire microbial genomes has led to the development of *synthetic biology*, a new field in which it is possible to ask, "How few parts (i.e., proteins and RNAs) does it take to construct a cell?" Recent experimental studies involving *Mycoplasma genitalium* (an obligate parasite requiring relatively little adaptive capability) suggest that the answer is 43 RNA-encoding and 400 or fewer protein-coding genes (Glass et al. 2006; Koonin 2000). The genomes of known viroids are approximately 1,000–2,000 times smaller than those of mycoplasmas, ranging in size from 246 to 401 nt. Has the time now come to ask, "What is the minimal size of the viroid genome?" Perhaps so, for much of the necessary preliminary information is either already in hand or is currently the subject of active investigation.

The existence of five structural domains in members of the *Pospiviroidae* was first proposed more than 20 years ago on the basis of the comparison of only eight different viroids (Keese and Symons 1985), and much effort has since been devoted to probing the biological role(s) of individual domains by a variety of techniques. The Subviral RNA Database now contains the complete sequences of more than 1,100 viroid variants, information that can be used to compare the evolutionary constraints operating on viroids. A series of RNA structural studies carried out by Detlev Riesner and colleagues and largely focused on PSTVd have identified several alternative interactions such as secondary hairpin II that are essential for viability (reviewed in Steger and Riesner 2003). Multimeric PSTVd RNAs undergo a complicated series of structural rearrangements during conversion into monomeric progeny but, a recent study by Schrader et al. (2003) clearly showed that a "miniRNA" containing only sequences from the central domain of PSTVd can be efficiently processed in a nuclear extract. Interactions with host protein(s) are almost certainly required for other steps in the replication pathway as well. For example, interaction of PSTVd with Virp1, an RNA-binding protein from tomato containing a nuclear localization signal, involves a discrete sequence/structural motif in the right terminal loop (Martínez de Alba et al. 2003). Following transport into the nucleus, synthesis of PSTVd (–) strands by RNA polymerase II initiates in the left terminal loop at either position 359 or position 1 (Kolonko et al. 2006).

Several studies have described the expression of potentially infectious viroid RNAs from transgenes inserted into the nuclear genomes of both host and nonhost plant species. This is a particularly powerful experimental strategy, because it allows time for an initial variant that is only very weakly infectious to mutate and evolve in vivo. As described earlier, Wassenegger and colleagues have used this strategy to demonstrate (1) a role for the loop E motif in determining PSTVd host range as well as (2) the evolution of a PSTVd variant lacking a large portion of the pathogencity domain. Such an error-correction mechanism will be very important in future efforts to design a minimal viroid genome capable of autonomous replication, because much about viroid–host interaction remains only poorly understood. Viroid replication appears to be relatively resistant to RNA silencing (Wang et al. 2004) but unanticipated interactions between individual structural domains provide a particular challenge. Sequence changes in all but the terminal right domain have been shown to influence symptom development (Sano et al. 1992; Qi and Ding 2003), and a series of five sequence changes in pathogenicity and variable domains of PSTVd disrupts transport from the bundle sheath to the leaf mesophyll (Qi et al. 2004). Even small deletions are likely to have unpredictable effects.

References

Ambrós S, Hernandez C, Desvignes JC, Flores R (1998) Genomic structure of three phenotypically different isolates of peach latent mosaic viroid: implications of the existence of constraints limiting the heterogeneity of viroid quasispecies. J Virol 72:7397–7406

Ambrós S, Hernandez C, Flores R (1999) Rapid generation of genetic heterogeity in progenies from individual cDNA clones of peach latent mosaic viroid in its natural host. J Gen Virol 80:2239–2252

Baumstark T, Schröder ARW, Riesner D (1997) Viroid processing: switch from cleavage to ligation is driven by a change from a tetraloop to a loop E conformation. EMBO J 16:599–610

Behjatnia SAA, Dry IB, Krake LR, Condé BD, Connelly MI, Randles JW, Rezaian MA (1996) New potato spindle tuber viroid and tomato leaf curl geminivirus strains from a wild *Solanum* sp. Phytopathology 86:880–886

Branch AD, Robertson HD (1984) A replication cycle for viroids and other small infectious RNA's. Science 223:450–455

Branch AD, Benenfeld BJ, Baroudy B.M, Wells FV, Gerin JL, Robertson HD (1985) Ultraviolet light-induced cross-linking reveals a unique region of local tertiary structure in potato spindle tuber viroid and HeLa 5S RNA. Proc Natl Acad Sci USA 82:6590–6594

Burke JM (2002) Hairpin and hammerhead ribozymes: how different are they? Biochem Soc Trans 30:1115–1118

Bussière F, Ouellet J, Côté F, Lévesque D, Perreault JP (2000) Mapping in solution shows the peach latent mosaic viroid to possess a new pseudoknot in a complex, branched secondary structure. J Virol 74:2647–2654

Codoñer FM, Daròs J-A, Solé RV, Elena SF (2006) The fittest versus the flattest: Experimental confirmation of the quasispecies effect with subviral pathogens. PLoS Pathog 2:1187–1193

Coté F, Levesque D, Perreault JP (2001) Natural 2′,5′-phosphodiester bonds found at the ligation sites of peach latent mosaic viroid. J Virol 75:19–25

Daros JA, Marcos JF, Hernandez C, Flores R (1994) Replication of avocado sunblotch viroid: evidence for a symmetric pathway with two rolling circles and hammerhead ribozyme processing. Proc Natl Acad Sci USA 91:12813–12817

Daros J-A, Elena SF, Flores R (2006) Viroids: an Ariadne's thread into the RNA labyrinth. EMBO Rep 7:593–588

Dickson E, Robertson HD, Niblett CL, Horst RK, Zaitlin M (1979) Minor differences between nucleotide sequences of mild and severe strains of potato spindle tuber viroid. Nature 277:60–62

Diener TO (1971) Potato spindle tuber "virus" IV. A replicating, low molecular weight RNA. Virology 45:411–428

Diener TO (1989) Circular RNAs: relics of precellular evolution? Proc Natl Acad Sci USA 86:9370–9374

Diener TO (1996) Origin and evolution of viroids and viroid-like satellite RNAs. Virus Genes 11:119–131

Diener TO (2001) The viroid: biological oddity or evolutionary fossil? Adv Virus Res 57:137–184

Diener TO (2003) Discovering viroids – a personal perspective. Nat Rev Microbiol 1:75–80

Ding B, Itaya A (2007) Viroid: a useful model for studying the basic principles of infection and RNA biology. Mol Plant Microbe Interact 20:7–20

Ding B, Itaya A, Zhong XH (2005) Viroid trafficking: a small RNA makes a big move. Curr Opin Plant Biol 8:606–612

Elena SF, Dopazo J, Flores R, Diener T, Moya A (1991) Phylogeny of viroids, viroidlike satellite RNAs, and the viroidlike domain of hepatitis δ virus RNA. Proc Natl Acad Sci USA 88:5361–5364

Elena SF, Dopazo J, de la Peña M, Flores R, Diener TO, Moya A (2001) Phylogenetic analysis of viroid and viroid-like satellite RNAs from plants: a reassessment. J Mol Evol 53:155–159

Flores R, Hernandez C, Martinez de Alba AE, Daros J Di Serio F (2005a) Viroids and viroid-host interactions. Annu Rev Phytopathol 43:117–139

Flores R, Randles JW, Owens RA, Bar-Joseph M, Diener TO (2005b) Viroids. In: Fauquet CM, Mayo MA, Maniloff J, Desselberger U, Ball AL (eds) Virus taxonomy. VIIII report of the International Committee on Taxonomy of Viruses. Elsevier/Academic, London, pp 1145–1159

Glass JI, Assad-Garcia N, Alperovich N, Yooseph S, Lewis MR, Maruf M, Hutchison CA 3rd, Smith HO, Venter JC (2006) Essential genes of a minimal bacterium. Proc Natl Acad Sci USA 103:425–430

Góra A, Candresse T, Zagórski W (1994) Analysis of the population structure of three phenotypically different PSTVd isolates. Arch Virol 138:223–245

Góra-Sochacka A, Kierzek A, Candresse T, Zágorski W (1997) The genetic stability of potato spindle tuber viroid (PSTVd) molecular variants. RNA 3:68–74

Gozmanova M, Denti MA, Minkov IN, Tsagris M, Tabler, M (2003) Characterization of the RNA motif responsible for the specific interaction of potato spindle tuber viroid RNA (PSTVd) and the tomato protein Virp1. Nucleic Acids Res 31:5534–5543

Gross HJ, Domdey H, Lossow C, Jank P, Raba M, Alberty H, Sänger HL (1978) Nucleotide sequence and secondary structure of potato spindle tuber viroid. Nature 273:203–208

Gruner R, Fels A, Qu F, Zimmat R, Steger G, Riesner D (1995) Interdependence of pathogenicity and replicability with potato spindle tuber viroid. Virology 209:60–69

Hadidi A, Flores R, Randles JW, Semancik JS (eds) (2003) Viroids. CSIRO, Collingwood

Hammond RW (1994) *Agrobacterium*-mediated inoculation of PSTVd cDNAs onto tomato reveals the biological effect of apparently lethal mutations. Virology 201:36–45

Hammond RW, Owens RA (1987) Mutational analysis of potato spindle tuber viroid reveals complex relationships between structure and infectivity. Proc Natl Acad Sci USA 84:3967–3971

Hammond R, Smith DR, Diener TO (1989) Nucleotide sequence and proposed secondary structure of Columnea latent viroid: a natural mosaic of viroid sequences. Nucleic Acids Res 17:10083–10094

Hampell KJ, Burke JM (2001) A conformational change in the "loop E-like" motif of the hairpin ribozyme is coincidental with domain docking and is essential for catalysis. Biochemistry 40:3723–3729

Haseloff J, Mohamed NA, Symon RH (1982) Viroid RNAs of cadang-cadang disease of coconuts. Nature 299:316–321

Hassen IF, Massart S, Motard J, Roussel S, Kummert J, Fakhfakh H, Marrakchi M, Perreault JP, Jijakli MH (2007) Molecular features of new peach latent mosaic viroid variants suggest that recombination may have contributed to the evolution of this infectious RNA. Virology 360:50–57

Hendrix DK, Brenner SE, Holbrook SR (2005) RNA structural motifs: building blocks of a modular biomolecule. Q Rev Biophys 38:221–243

Jenkins GM, Woelk CH, Rambaut A, Homes EC (2000) Testing the extent of sequence similarity among viroids, satellite RNAs, and the hepatitis delta virus. J Mol Evol 50:98–102

Juhasz A, Hegyi H, Solymosy F (1988) A novel aspect of the information content of viroids. FEBS Biochim Biophys Acta 950:455–458

Keese P, Symons RH (1985) Domains in viroids: evidence of intermolecular RNA rearrangements and their contribution to viroid evolution. Proc Natl Acad Sci USA 82:4582–4586

Kofalvi SA, Marcos JF, Canizares MC, Pallas V, Candresse T (1997) Hop stunt viroid (HSVd) sequence variants from *Prunus* species: evidence for recombination between HSVd isolates. J Gen Virol 78:3177–3186

Kolonko N, Bannach O, Aschermann K, Hu KH, Moors M, Schmitz M, Steger G, Riesner D (2006) Transcription of potato spindle tuber viroid by RNA polymerase II starts in the left terminal loop. Virology 347:392–404

Koltunow AM, Rezaian MA (1988) Grapevine yellow speckle viroid: structural features of a new viroid group. Nucleic Acids Res 16:849–864

Koonin EV (2000) How many genes can make a cell: the minimal-gene-set concept. Annu Rev Genomics Hum Genet 1:99–116

Leontis NB, Westhof E (1998) A common motif organizes the structure of multi-helix loops in 16S and 23 S ribosomal RNAs. J Mol Biol 283:571–583

Malfitano M, Di Serio F, Covelli L, Ragozzino A, Hernández C, Flores R (2003) Peach latent mosaic viroid variants inducing peach calico (extreme chlorosis) contain a characteristic insertion that is responsible for this symptomatology. Virology 313:492–501

Martínez de Alba AE, Sägasser R, Tabler M, Tsagris M (2003) A bromodomain-containing protein from tomato specifically binds potato spindle tuber viroid RNA *in vitro* and *in vivo*. J Virol 77:9685–9694

Martínez-Soriano JP, Galindo-Alonso J, Maroon CJ, Yucel I, Smith DR, Diener TO (1996) Mexican papita viroid: putative ancestor of crop viroids. Proc Natl Acad Sci USA 93:9397–9401

Matoušek J (2003) Hop latent viroid (HLVd) microevolution: an experimental transmission of HLVd "thermomutants" to solanaceous species. Biol Plant 46:607–610

Matoušek J, Patzak J, Orctova L, Schubert J, Vrba L, Steger G, Riesner D (2001) The variability of hop latent viroid as induced upon heat treatment. Virology 287:349–358

Matoušek J, Orctova L, Steger G, Skopek J, Moors M, Dedic P, Riesner D (2004) Analysis of thermal stress-mediated PSTVd variation and biolistic inoculation of progeny of viroid "thermomutants" to tomato and *Brassica* species. Virology 323:9–23

Mink GI, Wample R, Howell WE (1998) Heat treatment of perennial plants to eliminate phytoplasmas, viruses and viroids while maintaining plant survival. Breeding for resistance to plant viruses. In: Hadidi A, Kheterpal RK, Koganezawa H (eds) Plant virus disease control. APS, St Paul, pp 332–345

Navarro JA, Vera A, Flores R (2000) A chloroplastic RNA polymerase resistant to tagetitoxin is involved in replication of avocado sunblotch viroid. Virology 268:218–225

Owens RA, Yang G, Gundersen-Rindal D, Hammond RW, Candresse T, Bar-Joseph M (2000) Both point mutation and RNA recombination contribute to the sequence diversity of citrus viroid III. Virus Genes 20:243–252

Owens RA, Thompson SM, Kramer M (2003) Identification of neutral mutants surrounding two naturally-occurring variants of *Potato spindle tuber viroid*. J Gen Virol 84:751–756

Palacio-Bielsa A, Romero-Durban J, Duran-Vila N (2004) Characterization of citrus HSVd isolates. Arch Virol 149:537–552

Pelchat M, Levesque D, Ouellet J, Laurendeau S, Levesque S, Lehoux J, Thompson DA, Eastwell KC, Skreczkowski LJ, Perrault JP (2000) Sequencing of peach latent mosaic viroid variants from nine North American peach cultivars shows that this RNA folds into a complex secondary structure. Virology 271:37–45

Polivka H, Staub U, Gross HJ (1996) Variation of viroid profiles in individual grapevine plants: novel grapevine yellow speckle viroid 1 mutants show alterations of hairpin I. J Gen Virol 77:155–161

Puchta H, Herold T, Verhoeven K, Roenhorst A, Ramm K, Schmidt-Puchta W, Sänger HL (1990) A new strain of potato spindle tuber viroid (PSTVd-N) exhibits major sequences differences as compared to all other PSTVd strains sequenced so far. Plant Mol Biol 15:509–511

Qi Y, Ding B (2003) Inhibition of cell growth and shoot development by a specific nucleotide sequence in a noncoding viroid RNA. Plant Cell 15:1360–1374

Qi Y, Pélissier T, Itaya A, Hunt E, Wassenegger M, Ding B (2004) Direct role of a viroid RNA motif in mediating directional RNA trafficking across a specific cellular boundary. Plant Cell 16:1741–1752

Rakowski AG, Symons RH (1989) Comparative sequence studies of variants of avocado sunblotch viroid. Virology 173:352–356

Rigden JE, Rezaian MA (1993) Analysis of sequence variation in grapevine yellow speckle viroid 1 reveals two distinct alternative structures for the pathogenic domain. Virology 193:474–477

Sanjuán R, Forment J, Elena SF (2006) *In silico* predicted robustness of viroids (sic) RNA secondary structures. I. The effect of single mutations. Mol Biol Evol 23:1427–1436

Sano T (2003) Hop stunt viroid. In: Hadidi A, Flores R, Randles JW, Semancik JS (eds) Viroids. CSIRO, Collingwood, pp 207–212

Sano T, Candresse T, Hammond RW, Diener TO, Owens RA (1992) Identification of multiple structural domains regulating viroid pathogenicity. Proc Natl Acad Sci USA 89:10104–10108

Sano T, Mimura R, Ohshima K (2001) Phylogenetic analysis of hop and grapevine isolates of hop stunt viroid supports a grapevine origin for hop stunt disease. Virus Genes 22:53–59

Sano T, Yoshida H, Goshono M, Monma T, Kawasaki H, Ishizaki K (2004) Characterization of a new viroid strain from hops: evidence for viroid speciation by isolation in different host species. J Gen Plant Pathol 70:181–187

Schindler I-M, Mühlbach H-P (1992) Involvement of nuclear DNA-dependent RNA polymerases in potato spindle tuber viroid replication: a re-evaluation. Plant Sci 84:221–229

Schnell RJ, Olano CT, Kuhn DN (2001) Detection of avocado sunblotch viroid variants using fluorescent single-strand conformation polymorphism analysis. Electrophoresis 22:427–432

Schrader O, Baumstark T, Riesner D (2003) A mini-RNA containing the tetraloop, wobble-pair and loop E motifs of the central conserved region of potato spindle tuber viroid is processed into a minicircle. Nucleic Acids Res 31:988–998

Schultz ES, Folsom D (1923) Transmission, variation, and control of certain degeneration diseases of Irish potatoes. J Agric Res 25:43–118

Schuster P (2001) Evolution *in silico* and *in vitro*: the RNA model. Biol Chem 382:1301–1314

Semancik JS, Szychowski JA, Rakowski AG, Symons RH (1993) Isolates of citrus exocortis viroid recovered by host and tissue selection. J Gen Virol 74:2427–2436

Szychowski JA, Vidalakis G, Semancik JS (2005) Host-directed processing of *Citrus exocortis viroid*. J Gen Virol 86:473–477

Steger G, Riesner D (2003) Molecular characteristics. In: Hadidi A, Flores R, Randles JW, Semancik JS (eds) Viroids. CSIRO, Collingwood, pp 15–29

Tabler M, Tsagris M (2004) Viroids: petite RNA pathogens with distinguished talents. Trends Plant Sci 9:339–348

Verhoeven JTJ, Jansen C, Willemen TM, Kox LFF, Owens RA, Roenhorst JW (2004) Natural infections of tomato by *Citrus exocortis viroid, Columnea latent viroid, Potato spindle tuber viroid* and *Tomato chlorotic dwarf viroid*. Eur J Plant Pathol 110:823–831

Visvader JE, Symons RH (1985) Eleven new sequence variants of citrus exocortis viroid and the correlation of sequence with pathogencity. Nucleic Acids Res 13:2907–2920

Wang MB, Bian XY, Wu LM, Liu LX, Smith NA, Isenegger D, Wu RM, Masuta C, Vance VB, Watson JM, Rezaian A, Dennis ES, Waterhouse PM (2004) On the role of RNA silencing in the pathogenicity and evolution of viroids and viral satellites. Proc Natl Acad Sci USA 101:3275–3280

Warrilow D, Symons R (1999) Citrus exocortis viroid RNA is associated with the largest subunit of RNA polymerase II in tomato in vivo. Arch Virol 144:2367–2375

Wassenegger M, Heimes S, Sänger HL (1994) An infectious RNA replicon evolved from an *in vitro*-generated non-infectious viroid deletion mutant via a complementary deletion *in vivo*. EMBO J 13:6172–6177

Wassenegger M, Spieker RL, Thalmeir S, Gast FU, Riedel L, Sänger HL (1996) A single nucleotide substitution converts potato spindle tuber viroid (PSTVd) from a noninfectious to an infectious RNA for *Nicotiana tabacum*. Virology 226:191–197

Zhong X, Leontis N, Qian S, Itaya A, Qi Y, Boris-Lawrie K, Ding B (2006) Tertiary structural and functional analyses of a viroid RNA motif by isostericity matrix and mutagenesis reveal its essential role in replication. J Virol 80:8566–8581

Zhu Y, Qi Y., Xun R, Owens R, Ding B (2002) Movement of potato spindle tuber viriod reveals regulatory points of phloem-mediated RNA traffic. Plant Physiol 130:138–146

Chapter 6
Virus Populations, Mutation Rates and Frequencies

Justin S. Pita(✉) and Marilyn J. Roossinck

Abstract Populations of plant viruses are genetically heterogeneous. This heterogeneity is often linked to mutation, the ultimate source of genetic variation and an uncontested player in plant virus evolution. This review gives basic key information indispensable to understanding mutation in plant viruses, from mutation sources, mutation detection means, to the role of mutation in shaping plant virus evolution in combination with various other evolutionary factors. From information drawn from the recent literature, we confirm or refute some generally held views and we reinstate several unanswered questions. It is clear that low genetic diversity characterizes some plant virus populations, irrespective of their life cycle or their nature (DNA or RNA). Mutation frequencies of plant DNA viruses can be as high as those of RNA viruses. This casts some doubt on a positive correlation between high mutation rates and adaptive evolution, and on the lack of proofreading for RNA-dependent RNA polymerases. However, the lack of information on viral mutation rates still precludes a complete understanding of the link between mutation rates and population heterogeneity. Information about plant virus replication mode, generation time and gencration size also is still crucially needed before a complete picture of virus evolution will emerge.

Justin S. Pita
The S.R. Noble Foundation, P.O. Box 2180, Ardmore, OK 73402, USA
jspita@noble.org

M.J. Roossinck (ed.), *Plant Virus Evolution.*

6.1 Introduction

Variation is a characteristic of all living entities. Plant virus variation was reported as early as 1926 (Kunkel 1947) from observation of different symptom variants from the same virus source. Plant viruses develop genetic variation by errors occurring during the replication of their genomes and by reassortment. The two main types of errors are mutation (base substitution, insertion, deletion, inversion) and recombination. Recombination is a key mechanism for generating novel virus forms; however, recombination will not be addressed here (see Chap. 8). Instead, this chapter will review the literature pertaining to mutation and it will focus on rates and frequencies of point mutations.

Mutation is an error during DNA or RNA replication that results in a change in the sequence of nucleotide bases. Point mutation refers to a change in a single base of DNA or RNA sequence. Mutation rate and mutation frequency are two terms that are commonly used but frequently misunderstood. For the sake of clarity and to avoid confusion to the reader, we will clearly define these terms. Mutation rate refers to the polymerase fidelity or to the rate of nucleotide misincorporation by polymerase error, whereas mutation frequency refers only to the detectable occurrence of mutations after natural selection and genetic drift have acted on the mutant cloud produced by the mutation rate (Domingo and Holland 1994). As defined, mutation frequency is related to mutation rate by several variables, including selection, the mode of replication and the life history of the virus, and can differ from mutation rates by large factors (Drake 1993; Drake et al. 1998). Mutation frequencies are well studied for plants viruses, whereas little work has been done regarding mutation rates in plant virology. Consequently, this review will draw from studies in animal virology to evaluate the relevance of mutation frequencies and mutation rates to better our understanding of plant virus evolution.

6.2 Mutation Sources and Mutagenic Agents

Mutations result from several biological processes. In addition to misincorporation, mutations can be introduced by replication slippage, polymerase pausing, RNA splicing, RNA editing, oxidative damage and posttranscriptional modification. Replication slippage, also known as copy choice recombination, is a process that occurs between repeated sequences in both prokaryotes and eukaryotes and causes several human diseases (Hancock and Santibáñez-Koref 1998). Replication slippage takes place in vivo in *Escherichia coli*, between short direct repeats and longer tandem repeats (Bierne et al. 1997). Although slippage has not been directly demonstrated for plant viruses, the presence of short repeats in potyvirus sequences suggests that it may be involved in the evolution of theses viruses (Hancock et al. 1995). Similarly, it was proposed that polymerase pausing induces deletions and imprecise homologous recombination in the genome of a plant RNA virus (Nagy and Bujarski 1996). In DNA, a connection between replication slippage and polymerase pausing was established (Viguera et al. 2001).

RNA splicing and RNA editing alter the sequence of an RNA from that encoded in the DNA (Bass 2002). Unlike RNA splicing, which removes a large block of contiguous sequences, the RNA editing process changes one or two nucleotides. RNA splicing is an important process in the evolution of geminiviruses (Schalk et al. 1989) and caulimoviruses (Scholthof et al. 1991; Pennington and Melcher 1993). RNA editing is a common process in animal viruses (Polson et al. 1996; Cheng and Nagy 2003). Although it has not been described in a plant virus, a coordination between splicing and editing is very suggestive (Bass 2002). Moreover, plant cells utilize RNA editing mechanisms and contain the appropriate enzymes (Scott 1995).

Oxidative damage is a common process in DNA that results in mutations. Such damage is believed to be the root cause of many diseases, including cancer, heart disease and aging. Its mutagenic action affects cellular DNA and could mutagenize integrated provirus and replicating RNA genomes.

Gierer and Mundry (1958) showed that nitrous acid treatment of *Tobacco mosaic virus* (TMV) increased the spontaneous frequency of necrotic lesion mutants more than 20-fold. A variety of other chemicals can cause significant increases in mutation frequency in a wide variety of RNA virus genomes (Fields and Joklik 1969; Pringle 1970). Although Halle (1968) reported up to 220-fold increase in the frequency of large-plaque mutants of *Venezuelan equine encephalitis virus* by mutagenesis with 5-azacytidine, all viruses do not tolerate an increase in mutation rates. Indeed, some viral populations are extremely vulnerable to an increase in mutation rates. For example, small increases of the mutation rates of *Vesicular stomatitis virus* (VSV) and poliovirus by chemical mutagenesis were counterbalanced by a decrease in virus yield (lethality) (Holland et al. 1990), indicating the existence of a negative correlation between an increase of mutation rates and viability for certain viruses, for which a slight increase of mutation rate can lead to extinction. The correlation between high mutation rates and viral adaptation is discussed in Sec. 6.5.

Many other mutagenic agents exist, including ultraviolet radiation, a powerful agent that is widely used in mutagenesis studies in numerous organisms, in addition to some less common mutagens such as RNA or DNA secondary structures and host or environmental factors. RNA secondary structure was shown to trigger a threefold increase in mutation rates of a retrovirus (Pathak and Temin 1992). RNA secondary structure affects the deletion and insertion rates of the polymerase of *Cucumber mosaic virus* (CMV), as shown by simultaneously measuring the mutation rate in a structured and a nonstructured region of a satellite RNA reporter (Pita et al. 2007). In this study all the mutations detected were located within the structured region, which in this case consisted of a strong secondary structure of nine helices (Rodríguez-Alvarado and Roossinck 1997). On the other hand, the long single-stranded nonstructured region of the satellite reporter was without mutations, providing compelling evidence of the mutagenesis effect of the secondary structure.

Different mutation frequencies and different mutation rates were obtained for the same virus, CMV, in two different hosts, pepper and tobacco (Schneider and Roossinck 2001; Pita et al. 2007). This could reflect differences in the host factor(s) directly involved as part of the replicase, or differences in nucleotide pools or soluble components such as magnesium ions. Such differences in mutation rates in response to an environment have been well described for bacterial cultures exposed to growth-limiting stress, for instance, starvation (Rosenberg 2001). The process is termed stress-inducible mutation and the elements responsible for this type of mutation are called environment-dependent mutators. Mutagenic factors such as these are yet to be described for plant viruses.

6.3 Quantifying Methods

Different approaches are used to estimate mutation rates and mutation frequencies of plant viruses. Among them, molecular techniques are the most used at present. However, the choice of a given technique should be driven by the goal of the analysis. Some techniques provide only qualitative data and can be used to identify variants in a viral population. Other techniques can be used to quantify the extent of difference between identified variants. Qualitative techniques include ribonuclease protection assay (RPA; Aranda et al. 1995), restriction enzyme length polymorphism (RFLP; Naraghi-Arani et al. 2001) and single-stranded conformational polymorphism (SSCP; Koenig et al. 1995; Sanchez-Campos et al. 2002). These techniques do not allow a direct estimate of the diversity. In SSCP, PCR products rather than cloned fragments are tested, thus eliminating any concern of in vitro misincorporation by the polymerase as is the case during sequence analysis of cloned progeny. But the RPA, RFLP and SSCP techniques yield results that depend on the sequence context and cannot detect mutations that are not fixed in the population. Therefore, analysis of the nucleotide sequences of viral genes will yield the most detailed information, provided a great deal of care is taken to eliminate any in vitro mutation background during the reverse transcription, PCR and cloning processes (Schneider and Roossinck 2000). Sequence analysis of cloned progeny has been used to estimate the diversity of a number of plant viruses in experimental and natural population studies (reviewed in Roossinck and Schneider 2005, Tables 1, 2).

Although mutation frequencies are among the more directly measurable population parameters, the information needed to convert them into mutation rates is lacking because nothing is known about the mode of replication of plant viruses or generation times and generation sizes in plant viruses (Roossinck and Schneider 2005). Estimates have been made for TMV polymerase mutation rates (Malpica et al. 2002), and a direct measure of CMV polymerase indel rates has just been published (Pita et al. 2007).

6.4 Mutational Spectrum

To make up for the lack of data for the mutational spectrum for riboviruses, Drake and Holland (1999) estimated their spontaneous mutation rates using a correction factor derived from DNA-based microbes. A decade later, the first mutational spectrum of a ribovirus was shown to be very different from those of most DNA-based organisms (Malpica et al. 2002). It contains the common mixture of base-pair substitutions and insertions and deletions (indels). However, the base-pair substitutions are in a minority (11/35), presenting a different view for ribovirus mutation and evolution. Such a ratio of base-pair substitutions to indels had only been reported for a retrovirus and for an archeon (Pathak and Temin 1990; Grogan et al. 2001). Most indels are deleterious mutations because they disrupt open reading frames. Also, because of the role of RNA secondary structure in ribovirus life history, even synonymous mutations can be deleterious. Consequently, the large fraction (69%) of indels within the TMV mutational spectrum (Malpica et al. 2002) indicates a very high rate of deleterious mutations within TMV populations. This suggests that high mutation rates of this virus may not necessarily be adaptive. However, although indels are usually lethal, they also create the raw material for evolutionary leaps, by creating new open reading frames, and the potential for expression of entirely novel proteins.

Recently, we have shown a disproportionate ratio of one insertion to 49 deletions in the estimates for CMV indel rate *in planta*. This difference is not reflected in the recovered incidence of indels after extended infection of CMV in plants (Pita et al. 2007), but this huge difference in deletion and insertion rates may explain the minimal length of most extant RNA viruses and emphasizes the strength of selection in maintaining virus genomes because of the rarity of reversion for deletion mutations.

6.5 Mutation Rates Versus Adaptation

Earlier experiments in virus evolution studies involved a host shift whereby a virus adapted to a particular host was passaged into a different one. The resulting change of virus traits was called host adaptation (Yarwood 1979). Virus mutation is the ultimate source of the genetic variation. Because the genetic variation is required for adaptation, it is tempting to think that the genetic mutation rates would be tuned to a level that best promotes adaptation. This means that the mutation rate would always be at its highest level in order to speed up the response to any selection pressure. We saw that this is not always the case with the example of TMV for which 69% of the mutations are deleterious (Malpica et al. 2002). Indeed it is likely that selection to decrease mutation rates was the reason to use DNA instead of RNA as the hereditary molecule (Maynard and Szathmary 1995). On the other hand, a deleterious-compensatory evolution model whereby the fixation of a deleterious mutation can trigger subsequent positive selection of compensatory mutations was proposed for viruses of the *Potyviridae*, the largest family of plant viruses (Wang et al. 2006). Considering this deleterious-compensatory model and given the evidence that evolvability of an RNA virus depends on mutational neighborhood (Burch and Chao 2000), the high rate of deleterious mutations present in the TMV mutational spectrum still could have favored TMV's adaptation in a new niche. Recent studies have examined the effect of mutation rate on the speed of adaptive evolution. Furió et al. (2005) used VSV to demonstrate that there was no positive correlation between mutation and adaptation rates. They concluded that the mutation rate in VSV resulted from a trade-off between replication rate and replication fidelity. However, Vignuzzi et al. (2006) came to a different conclusion. They found that poliovirus replicating at high fidelity generates less genomic diversity and is unable to adapt to adverse growth conditions. Therefore, poliovirus adaptation can be limited by low mutation rates. Thus, the relationship between high ribovirus mutation rates and adaptive evolution remains an open question.

6.6 Intrahost Versus Interhost Diversity of Plant Virus Population

In spite of the positive correlation established between host ranges and mutation frequencies in experimental evolution studies (Schneider and Roossinck 2000, 2001), genetic stability is often the rule, rather than the exception, when comparing plant isolates (García-Arenal et al. 2003). Low genetic diversity characterizes many plant virus populations irrespective of their life cycle, or their nature (RNA or DNA) (reviewed in García-Arenal et al. 2001). The example of CMV, the plant RNA virus with the broadest host range of any known virus, confirms the rule. Its genetic diversity among isolates from California, estimated on the basis of haplotype frequencies and nucleotide distances for several genes, also appeared to be low (Lin et al. 2004). In experimental evolution studies looking at intrahost populations of CMV, consensus sequences did not change, even though populations had significant levels of mutation frequency (Schneider and Roossinck 2000, 2001). Similarly, the diversity of *Wheat streak mosaic virus*, another plant virus with a relatively broad host range, was low (French and Stenger 2003), and 17 isolates from Australia analyzed by consensus sequence were essentially identical to each other and to isolates from the USA (Dwyer et al. 2007). Recent analyses of the population structure of *Citrus psorosis virus* and *Barley yellow dwarf virus* have shown the same low genetic diversity among strains (Hall 2006; Martin et al. 2006). This probably results from the combined actions of natural selection (Domingo and Holland 1994) and genetic bottlenecks (Li and Roossinck 2004; Ali et al. 2006) on the high replication rates. However, some plant viruses do exhibit high levels of genetic diversity, both in individual populations and between isolates. *Banana mild mosaic virus*, a virus thought to be only transmitted vertically, shows high levels of diversity both within individual plants and between plants (Teycheney et al. 2005). This could be due to different selective constraints on this virus, compared with horizontally transmitted viruses.

A recent review summarizes the published data on plant virus mutation frequencies (Roossinck and Ali 2007). Unfortunately, almost all of the published data on plant virus diversity has been obtained on isolates from crop plants, which are in no way representative of what a plant virus would encounter in a natural setting. Another important factor for plant viral diversity is the mode of virus replication, which will ultimately dictate the rates at which mutations accumulate.

6.7 Replication Strategy

Information about the replication mode of RNA viruses would greatly help in understanding the uncorrelated relationship between high mutation rates and low virus diversity in isolates from crops. However, such information is still lacking. An RNA virus replication model is not yet well defined. It is not known if these viruses use the "stamping machine" mechanism (Luria and Delbruck 1943), an essentially linear replication model in which the parental molecule makes a single copy pregenome that is the only template used for production of the progeny, or if they utilize an exponential mechanism whereby the progenies of the initial molecule are able to become a template for further replication or if they use a mixture of both mechanisms

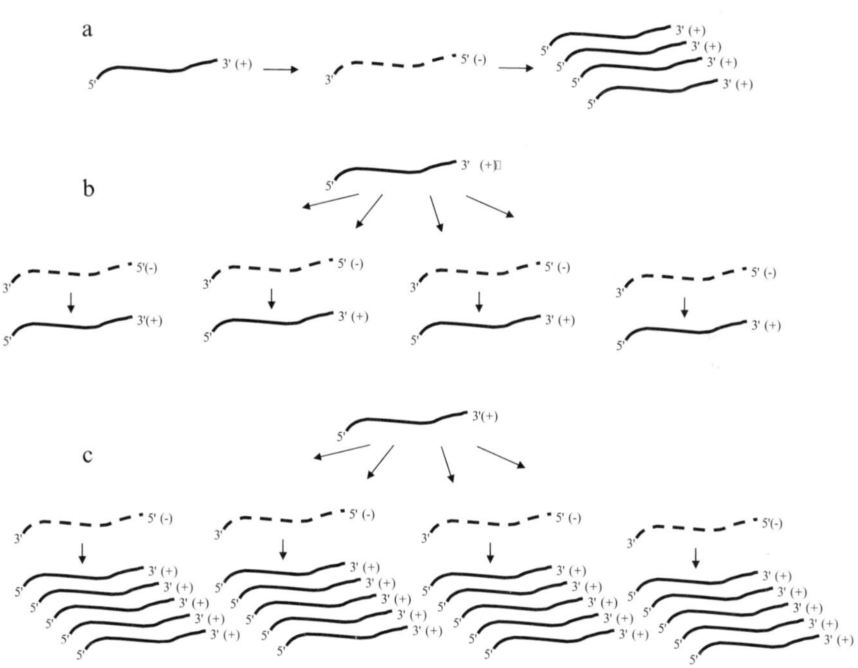

Fig. 6.1 Different potential modes of replication for a single-stranded virus. **a** The "stamping machine" model, where incoming (+) strand genomes are copied once to a (–) strand, which then serves as a template for all the progeny. **b** A modified model, where the incoming (+) strand generates a number of (–) strands, which in turn each generate a single (+) strand. **c** A fully exponential model, where the incoming (+) strand generates numerous (–) strands, each of which generates numerous (+) strands

(Fig. 6.1). A stamping machine model was proposed for the bacteriophage Φ6 (Chao et al. 2002), VSV was shown to replicate via an exponential mechanism (Cuevas et al. 2005) and *Herpes simplex virus type 1* (HSV-1), a human DNA virus, was proposed to possess a mixture of exponential and linear components (Drake and Hwang 2005), but for plant viruses the replication mode remains to be determined. It cannot be deduced by mathematical extrapolations not only because information about generation times is also lacking but also because amplification could occur at any stage in the replication cycle (Roossinck and Schneider 2005).

6.8 RNA Versus DNA Viruses

The best quality mutation rate data are those for animal DNA viruses. A compilation of these data indicated that all DNA-based microbes, in spite of their diverse genomes and irrespective of their life histories, showed a genomic rate of spontaneous mutation close to 0.003 (Drake 1991). The robustness of the standard genomic rate for DNA microbes was once more confirmed by the newest data added to the list. HSV-1, a human pathogenic DNA virus, has an estimated genomic mutation rate compatible with DNA values (Drake and Hwang 2005). There is no estimate for the mutation rate of plant DNA viruses to date, but mutation frequencies were measured for several plant DNA viruses. Because studies on populations of other plant DNA viruses are still scarce, we will focus on geminiviruses. They represent the best example of emerging viruses from the plant kingdom (see Chap. 3). Geminiviruses have a circular single-stranded DNA genome that uses the host plant DNA polymerase for replication (Bisaro 1996). High mutation frequencies were observed for *Maize streak virus*, genus *Mastrevirus* (Isnard et al. 1998), and for begomoviruses in wild and cultivated hosts (Ooi et al. 1997; Sanz et al. 1999). Mutation frequencies estimated for three open reading frames of *Cotton leaf curl virus* (CLCuV) exceeded the value reported for different genes in several plant and animal RNA viruses (Sanz et al. 1999). In spite of this, a low genetic diversity was observed among CLCuV isolates (Sanz et al. 1999). A temporal evolution study of *Tomato yellow leaf curl Sardinia virus* (TYLCSV) isolates (over 8 years) also showed that TYLCSV isolates have a low genetic diversity (Sanchez-Campos et al. 2002). Recently, experimental evolution studies of *Tomato yellow leaf curl china virus* demonstrated a quasispecies-like population structure, with mutation frequencies similar to those found in RNA virus experimental evolution (Ge et al. 2007). Such mutation frequency values, equivalent to those of RNA viruses, are surprising. A generally held view is that RNA viruses are prone to higher rates of mutation than DNA viruses because of their error-prone polymerases that lack proofreading capabilities (Steinhauer and Holland 1986; Domingo and Holland 1997). This view was challenged by some authors (Smith and Inglis 1987; Roossinck 1997) and reliable information on polymerase fidelity of RNA plant virus is still lacking. Geminiviruses lack genes for DNA polymerases. Their replication depends on host plant enzymes, and occurs in the host cell nuclei by a rolling-circle mechanism (Hanley-Bowdoin et al. 1999). Moreover, it was shown that the mutation repair mechanisms of the host probably do not function in the geminivirus cycle (Inamdar et al. 1992). This may explain their high mutation frequencies although some have argued that recombination is the major mechanism for geminivirus evolution (Martin et al. 2001; Pita et al. 2001; Ndunguru et al. 2005; Vadivukarasi et al. 2007).

6.9 Virus Populations

RNA viruses are capable of generating an enormous number of progeny and a high level of variation. In addition, selection often acts on the population as a whole, rather than on individual members of the population. Theoretical treatments of virus populations include the quasispecies theory, which has generated a great deal of discussion and controversy (Domingo 2002; Holmes and Moya 2002). While some have argued that quasispecies theory is unnecessary, and that viral populations can be described using standard population genetics, the concept is also often misunderstood. Quasispecies are defined as single replicating populations that are at an equilibrium of mutation and selection. Some of the misunderstandings arise because populations are probably sampled when they are not at equilibrium. In a quasispecies selection acts on the mutant spectrum rather than the individual genomes, so variants in a population can complement each other and result in extended function (reviewed in Bull et al. 2005).

6.10 Conclusion

A substantial amount of literature is available for plant virus mutation frequencies. On the other hand, information on mutation rates of plant viruses is crucially lacking. To make this review possible, we have drawn analogies from the literature existing in animal virology. However, because of significant differences between plant and animal viruses, more estimates of plant virus mutation rates are required to specifically understand the evolution of plant viruses. It is not yet possible to deduce the evolutionary trajectory of plant viruses from mutation frequencies because the information needed to estimate the mutation rate from mutation frequency is not available. Some very important questions pertaining to a better understanding of plant virus evolution are still unanswered. Information about plant virus generation time and generation size is lacking. The replication mode used by RNA viruses is still unknown. Consequently, some uncertainties remain regarding the grounds for the positive correlation that is often made between RNA virus high mutation rates and RNA virus potential for adaptive evolution. Moreover, mutation frequencies of some plant DNA viruses are equivalent to those of plant RNA viruses. This casts some doubt on the assumption that is often made between RNA virus high mutation rates and the lack of any repair system during RNA virus replication. Indeed, regardless of the type of the replicating virus (DNA or RNA), populations of plant viruses are often not highly variable and genetic stability seems to be more common than predicted.

Acknowledgement The authors were supported in full by the Samuel Roberts Noble Foundation.

References

Ali A, Li H, Schneider WL, Sherman DJ, Gray S, Smith D, Roossinck MJ (2006) Analysis of genetic bottlenecks during horizontal transmission of *Cucumber mosaic virus*. J Virol 80(17):8345–8350

Aranda MA, Fraile A, Garcia-Arenal F, Malpica J (1995) Experimental evaluation of the ribonuclease protection assay method for the assessment of genetic heterogeneity in populations of RNA viruses. Arch Virol 140:1373–1383

Bass BL (2002) RNA editing by adenosine deaminases that act on RNA. Annu Rev Biochem 71(1):817–846

Bierne H, Vilette D, Ehrlich S, Michel B (1997) Isolation of a *dnaE* mutation which enhances RecA-independent homologous recombination in the *Escherichia coli* chromosome. Mol Biol 24(6):1225–1234

Bisaro DM (1996) Geminivirus DNA replication. In: DePamphilis ML (ed) DNA Replication in eukaryotic cells. Cold Spring Harbor Laboratory Press, Cold Spring Harbor, pp 833–854

Bull JJ, Ancel ML, Lachmann M (2005) Quasispecies made simple. PLoS Biol 1(6):450–460

Burch CL, Chao L (2000) Evolvability of an RNA virus is determined by its mutational neighbourhood. Nature 406(10):625–628

Chao L, Rang CU, Wong, LE (2002) Distribution of spontaneous mutants and inferences about the replication mode of the RNA bacteriophage ø6. J Virol 76(7):3276–3281

Cheng C-P, Nagy PD (2003) Mechanism of RNA recombination in carmo- and tombusviruses: evidence for template switching by the RNA-dependent RNA polymerase in vitro. J Virol 77(22):12033–12047

Cuevas JM, Moya A, Sanjuan R (2005) Following the very initial growth of biological RNA viral clones. J Gen Virol 86(2):435–443

Domingo E (2002) Quasispecies theory in virology. J Virol 76(1):463–465

Domingo E Holland JJ (1994) Mutation rates and rapid evolution of RNA viruses. In: Morse SS (ed) The evolutionary biology of viruses. Raven, New York, pp 161–184

Domingo E, Holland JJ (1997) RNA virus mutations and fitness for survival. Annu Rev Microbiol 51:151–178

Drake JW (1991) A constant rate of spontaneous mutation in DNA-based microbes. Proc Natl Acad Sci USA 88:7160–7164

Drake JW (1993) Rates of spontaneous mutation among RNA viruses. Proc Natl Acad Sci USA 90:4171–4175

Drake JW, Holland JJ (1999) Mutation rates among RNA viruses. Proc Natl Acad Sci USA 96(24):13910–13913

Drake JW, Hwang CBC (2005) On the mutation rate of herpes simplex virus type I. Genetics 170:969–970

Drake JW, Charlesworth B, Charlesworth D, Crow JF (1998) Rates of spontaneous mutation. Genetics 148:1667–1686

Dwyer GI, Gibbs MJ, Gibbs AJ, Jones RAC (2007) *Wheat streak mosaic virus* in Australia: relationship to isolates from the Pacific northwest of the USA and its dispersion via seed transmission. Plant Dis 91(2):164–170

Fields BN, Joklik, WK (1969) Isolation and preliminary genetic and biochemical characterization of temperature-sensitive mutants of reovirus. Virology 37:335–342

French R, Stenger DC (2003) Evolution of Wheat streak mosaic virus: dynamics of population growth within plants may explain limited variation. Annu Rev Phytopathol 41:199–214

Furió V, Moya A, Sanjuan R (2005) The cost of replication fidelity in an RNA virus. Proc Natl Acad Sci USA 102(69):10233–10237

García-Arenal F, Fraile A, Malpica J (2003) Variation and evolution of plant virus populations. Int Microbiol 6:225–232

García-Arenal F, Fraile A, Malpica JM (2001) Variability and genetic structure of plant virus populations. Annu Rev Phytopathol 39:157–186

Ge L, Zhang J, Zhou X, Li H (2007) Genetic structure and population variability of tomato yellow leaf curl China virus. J Virol 81(11):5902–5907

Gierer A, Mundry K-W (1958) Production of mutants of tobacco mosaic virus by chemical alteration of its ribonucleic acid *in vitro*. Nature 182(4647):1457–1458

Grogan DW, Carver GT, Drake JW (2001) Genetic fidelity under harsh conditions: analysis of spontaneous mutation in the thermoacidophilic archaeon *Sulfolobus acidocaldarius*. Proc Natl Acad Sci USA 98(14):7928–7933

Hall G (2006) Selective constraint and genetic differentiation in geographically distant barley yellow dwarf virus populations. J Gen Virol 87(10):3067–3075

Halle S (1968) 5-Azacytidine as a mutagen for arboviruses. J Virol 2(10):1228–1229

Hancock JM, F Santibáñez-Koref MF (1998) Trinucleotide expansion disease in the context of micro- and minisatellite evolution, Hammersmith Hospital, April 1–3, 1998. EMBO J 17(19):5521–5524

Hancock JM, Chaleeprom W, Chaleeprom W, Dale J, Gibbs A (1995) Replication slippage in the evolution of potyviruses. J Gen Virol 76:3229–3232

Hanley-Bowdoin L, Settlage SB, Orozco BM, Nagar S Robertson D (1999) Geminiviruses: models for plant DNA replication, transcription, and cell cycle regulation. Crit Rev Plant Sci 18(1):71–106

Holland JJ, Domingo E, de la Torre JC, Steinhauer DA (1990) Mutation frequencies at defined single codon sites in vesicular stomatitis virus and poliovirus can be increased only slightly by chemical mutagenesis. J Virol 64(8):3960–3962

Holmes EC, Moya A (2002) Is the quasispecies concept relevant to RNA viruses? J Virol 76(1):460–462

Inamdar NM, Zhang X-Y, Brough CL, Gardiner WE, Bisaro DM, Ehrlich M (1992) Transfection of heteroduplexes containing uracil-guanine or thymine-guanine mispairs into plant cells. Plant Mol Biol 20:123–131

Isnard M, Granier M, Frutos R, Reynaud B, Peterschmitt M (1998) Quasispecies nature of three maize streak virus isolates obtained through different modes of selection from a population used to assess response to infection of maize cultivars. J Gen Virol 79:3091–3099

Koenig RP, Luddecke P, Haeberle AM (1995) Detection of been necrotic yellow vein virus strains, variants and mixed infections by examining single-stranded conformation polymorphisms on immunocaptured RT-PCR products. J Gen Virol 76:2051–2055

Kunkel LO (1947) Variation in phytopathogenic viruses. Annu Rev Microbiol 1:85–100

Li H, Roossinck, MJ (2004). Genetic bottlenecks reduce population variation in an experimental RNA virus population. J Virol 78(19):10582–10587

Lin H-X, Rubio L, Smythe AB, Falk BW (2004) Molecular population genetics of *Cucumber mosaic virus* in California: evidence for founder effects and reassortment. J Virol 78(12):6666–6675

Luria SE, Delbruck M (1943) Mutations of bacteria from virus sensitivity to virus resistance. Genetics 28:491–511

Malpica JM, Fraile A, Moreno I, Obies CI, Drake JW, García-Arenal F (2002) The rate and character of spontaneous mutations in an RNA virus. Genetics 162:1505–1511

Martin DP, Willment JA, Billharz R, Velders R, Odhiambo B, Njuguna J, James D, Rybicki EP (2001) Sequence diversity and virulence in *Zea mays* of *Maize streak virus* isolates. Virology 288(2):247–255

Martin S, Garcia ML, Troisi A, Rubio L, Legarreta G, Grau O, Alioto D, Moreno P, Guerri J (2006) Genetic variation of populations of *Citrus psorosis virus*. J Gen Virol 87(10):3097–3102

Maynard SJ, Szathmary E (1995) The major transitions in evolution. Freeman/Spektrum, Oxford
Nagy PD, Bujarski J (1996) Homologous RNA recombination in brome mosaic virus AU-rich sequences decrease the accuracy of crossovers. J Virol 70(1):415–426
Naraghi-Arani P, Daubert S, Rowhani A (2001) Quasispecies nature of the genome of *Grapevine fanleaf virus*. J Gen Virol 82:1791–1795
Ndunguru J, Legg J, Aveling T, Thompson G, Fauquet C (2005) Molecular biodiversity of cassava begomoviruses in Tanzania: evolution of cassava geminiviruses in Africa and evidence for East Africa being a center of diversity of cassava geminiviruses. Virol J 2(1):21
Ooi K, Ohshita S, Ishii I, Yahara T (1997) Molecular phylogeny of geminivirus infecting wild plants in Japan. J Plant Res 110(2):247–257
Pathak V, Temin H (1990) Broad spectrum of in vivo forward mutations, hypermutations, and mutational hotspots in a retroviral shuttle vector after a single replication cycle: substitutions, frameshifts, and hypermutations. Proc Natl Acad Sci 87(16):6019–6023
Pathak VK, Temin HM (1992) 5-Azacytidine and RNA secondary structure increase the retrovirus mutation rate. J Virol 66(5):3093–3100
Pennington RE, Melcher U (1993) *In planta* deletion of DNA inserts from the large intergenic region of Califlower mosaic virus DNA. Virology 192:188–196
Pita JS, de Miranda JR, Schneider WL, Roossinck MJ (2007) Environment determines fidelity for an RNA virus replicase. J Virol 81(17):9072–9077
Pita JS, Fondong VN, Sangaré A, Otim-Nape GW, Ogwal S, Fauquet CM (2001) Recombination, pseudorecombination and synergism of geminiviruses are determinant keys to the epidemic of severe cassava mosaic disease in Uganda. J Gen Virol 82:655–665
Polson AG, Bass BL, Casey JL (1996) RNA editing of hepatitis delta virus antigenome by dsRNA adenosine deaminase. Nature 380:454–456
Pringle CR (1970) Genetic characteristics of conditional lethal mutants of vesicular stomatitis virus induced by 5-fluorouracil, 5-azacytidine, and ethyl methane sulfonate. J Virol 5(5):559–567
Rodríguez-Alvarado G, Roossinck MJ (1997) Structural analysis of a necrogenic strain of cucumber mosaic cucumovirus satellite RNA *in planta*. Virology 236:155–166
Roossinck MJ (1997) Mechanisms of plant virus evolution. Ann Rev Phytopathol 35:191–209
Roossinck MJ, Ali A (2007) Mechanisms of plant virus evolution and identification of genetic bottlenecks: impact on disease management. In: Punja ZK, DeBoer SH, Sanfaçon H (eds) Biotechnology and plant disease management. CABI, Wallingford, pp 109–124
Roossinck MJ, Schneider WL (2005) Mutant clouds and occupation of sequence space in plant RNA viruses. In: Domingo E (ed) Quasispecies: concepts and implications for virology. Current topics in microbiology and immunology, vol 299. Springer, Heidelberg, pp 337–348
Rosenberg SM (2001) Evolving responsively: adaptive mutation. Nat Rev Genet 2:504–515
Sanchez-Campos S, Diaz JA, Monci F, Bejarano ER, Reina J, Vavas-Castillo MA, Aranda MA, Moriones E (2002) High genetic stability of the begomovirus *Tomato yellow leaf curl Sardinia* virus in southern Spain over an 8-year period. Phyopathology 92(8):842–849
Sanz AI, Fraile A, Gallego JM, Malpica JM, García-Arenal F (1999). Genetic variability of natural populations of Cotton leaf curl geminivirus, a single-stranded DNA virus. J Mol Evol 49(5):672–681
Schalk H-J, Matzeit V, Schiller B, Schell J, Gronenborn B (1989) Wheat dwarf virus, a geminivirus of graminaceous plants needs splicing for replication. EMBO J 8(2):359–364
Schneider WL, Roossinck MJ (2000) Evolutionarily related sindbis-like plant viruses maintain different levels of population diversity in a common host. J Virol 74(7):3130–3134
Schneider WL, Roossinck MJ (2001) Genetic diversity in RNA viral quasispecies is controlled by host-virus interactions. J Virol 75(14):6566–6571
Scholthof H, Wu FC, Richins RD, Shepherd RJ (1991) A naturally occurring deletion mutant of Figwort mosaic virus (caulimovirus) is generated by RNA splicing. Virology 184:290–298
Scott J (1995) A place in the world for RNA editing. Cell 81:833–836
Smith DB, Inglis SC (1987) The mutation rate and variability of eukaryotic viruses: an analytical review. J Gen Virol 68(11):2729–2740

Steinhauer DA, Holland JJ (1986) Direct method for quantification of extreme polymerase error frequencies at selected single base sites in viral RNA. J Virol 57:219–228

Teycheney P-Y, Laboureau N, Iskra-Caruana M-L, Candresse T (2005) High genetic variability and evidence for plant-to-plant transfer of *Banana mild mosaic virus*. J Gen Virol 86:3179–3187

Vadivukarasi T, Girish KR, Usha R (2007) Sequence and recombination analyses of the geminivirus replication initiator protein. J Biosci 32(1):17–29

Vignuzzi M, Stone JK, Arnold JJ, Cameron CE, Andino R (2006) Quasispecies diversity determines pathogenesis through cooperative interactions in a viral population. Nature 439:344–348

Viguera E, Canceill D, Ehrlich SD (2001) Replication slippage involves DNA polymerase pausing and dissociation. EMBO J 20(10):2587–2595

Wang H, Huang LF, Cooper JI (2006) Analyses on mutation patterns, detection of population bottlenecks, and suggestion of deleterious-compensatory evolution among members of the genus Potyvirus. Arch Virol 151(8):1625–1633

Yarwood CE (1979) Host passage effects with plant viruses. Adv Virus Res 24:69–90

Chapter 7
Genetic Bottlenecks

Akhtar Ali(✉) and Marilyn J. Roossinck

Abstract Changes in population size may have important effects on genetic variation and on the survival potential of viral species. Genetic bottlenecks are evolutionary events that reduce genetic variation of a population in a stochastic manner and result in founding populations that can lead to genetic drift. In nature, genetic bottlenecks may occur at different points during the life cycles of plant RNA viruses. For example, transmission events, both horizontal and vertical, and systemic infections represent events in the virus life cycles that may impose a bottleneck. Recently, genetic bottlenecks have been shown experimentally in plant virus populations during systemic movement within the plant and horizontal transmission from plant to plant by aphid vectors. The most important implication of genetic bottlenecks is the reduction in population size and intensification of genetic drift, which can reshape the RNA virus population and may lead to the emergence of new virus strains. Another effect of genetic bottlenecks is to reduce the size of the effective populations below the threshold needed to ensure the transmission of the fittest variants. Consequently, the viral population may become progressively dominated by less fit mutants, a process known as Muller's ratchet, and will succumb by a mutational meltdown.

Akhtar Ali, Ph.D.
Department of Biological Science, The University of Tulsa, 800 South Tucker Drive,
Tulsa, OK 74104-3189, USA
akhtar-ali@utulsa.edu

M.J. Roossinck (ed.), *Plant Virus Evolution.*

7.1 Introduction

Plant viral diseases constitute a major threat to the large-scale production of crops worldwide. Currently, more than 1,000 plant viruses have been reported and 80% of them have RNA as their genetic material (Fauquet et al. 2005). Plant viruses with RNA genomes are capable of generating highly polymorphic populations, sometimes referred to as quasispecies (see Chap. 6), that help them to overcome different selection pressures during the infection and invasion of different host plants. Understanding the variability of plant viruses is important because viral strains may differ in the severity of disease symptoms that they produce, in their ability to infect a given host plant, and in transmission, which ultimately can affect patterns of disease spread in the field. In animal viruses population diversity has been linked to both increased pathogenesis (Vignuzzi et al. 2006) and reduced virulence (Jerzak et al. 2007), depending on the virus.

High mutation rates, recombination, and reassortment are the three major mechanisms that enable plant viruses to produce highly diverse populations (Roossinck 1997). However, these three mechanisms are counterbalanced by forces that restrict the genetic variation of plant viruses in nature. One is negative selection, which specifically limits variation. The other is genetic bottlenecks, which randomly limit genetic variation of viral populations.

Genetic bottlenecks are evolutionary events that reduce genetic variation of a population in a stochastic manner and result in small founding populations that lead to genetic drift. Reduced genetic variation means that the population may not be able to adapt to new selection pressures, such as climate change or a change of host plant.

Genetic bottlenecks may occur at different points during the life cycles of RNA viruses. For example, systemic infection can pose a bottleneck when a plant virus moves from the initially infected tissue to invade other tissues. The majority of plant viruses are transmitted by vectors (most commonly insects) in the field to infect other hosts. Transmission can impose a bottleneck when a virus is transmitted from plant to plant with the help of their vectors. Some plant viruses are transmitted through seed, which can also impose a bottleneck when the virus moves from the vegetative tissues to reproductive tissues or seed. Genetic bottlenecks also occur when the plant population is reduced at the end of the growing season, or when the environment changes to limit infection of the plant or to inhibit virus reproduction (e.g., periods of hot, dry weather or a deep freeze).

The effects of artificial genetic bottlenecks on viral fitness have been studied extensively with animal and bacterial viruses (Chao 1990; Clarke et al. 1993; Novella et al. 1999; Yuste et al. 1999). However, only a few studies have looked at the effects of a natural bottleneck on a virus population. This chapter reviews current knowledge of the genetic bottlenecks for plant viruses and their effects on virus evolution.

7.2 Bottlenecks During Systemic Infections

Plant tissues contain various cell types that are connected to each other by plasmodesmata (Lucas 1999). After inoculation, plant viruses spread through plasmodesmata from the initially infected epidermal cells through the underlying mesophyll cells to adjacent cells (cell-to-cell movement) until they reach the phloem for rapid invasions of the younger plant parts (long-distance or vascular movement).

Systemic movement of a virus throughout its host plant is required for a complete infection of the host. Thus, the virus must be capable of moving from the initially infected cells to the vasculature, through which it is transported along with photoassimilates to distal parts of the plant. Once it arrives at a new tissue, the virus must exit the phloem and continue to move from cell to cell. The large numbers of plasmodesmata between mesophyll cells suggest that cell-to-cell movement probably does not represent a major bottleneck (Li and Roossinck 2004). It seems more likely that bottlenecks occur at the site of entry into or exit from the vasculature, where plasmodesmata are more limited and may be more restricted (Roossinck and Schneider 2005); thus, plant structures act as potential bottlenecks during the long-distance movement of RNA viruses within an infected plant. However, details of the dynamics of virus loading into and unloading from the phloem are not well understood. Plasmodesmatal structures vary among plant species (Turgeon 1996), and accumulation of detectable virus in phloem-associated cell types also varies among plants (Cillo et al. 2002a, b). It seems likely that bottlenecks in systemic infection also vary among plant hosts.

So far only three independent studies have shown experimentally that plant RNA viruses encounter severe genetic bottlenecks during systemic movement (Table 7.1). The three viruses – *Tobacco mosaic virus* (TMV; Sacristán et al. 2003), *Wheat streak mosaic virus* (WSMV; French and Stenger 2003), and *Cucumber mosaic virus* (CMV; Li and Roossinck 2004) – belong to three different genera and have very different host ranges. However, the estimates of the severity of bottlenecks during their systemic movement in the respective hosts were similar. Further studies are required with a number of other plant viruses during the systemic infection of various host species to assess the role of genetic bottlenecks.

Table 7.1 Genetic bottleneck studies during systemic movement of plant viruses

Virus genus	Virus name	Hosts used	Reference
Tritimovirus	*Wheat streak mosaic virus*	Monocotyledon	French and Stenger (2003)
Tobamovirus	*Tobacco mosaic virus*	Dicotyledon	Sacristan et al. (2003)
Cucumovirus	*Cucumber mosaic virus*	Dicotyledon	Li and Roossinck (2004)

7.3 Bottlenecks During Horizontal Transmission

The majority of plant viruses are transmitted by insect vectors in a nonpersistent, semipersistent, or persistent manner. Vector transmission of plant viruses was widely thought to be an important bottleneck (Pirone and Blanc 1996; Albiach-Marti et al. 2000) but there has been no direct experimental evidence to support this view until recently.

Recently genetic bottlenecks were demonstrated during the horizontal transmission of artificial populations of CMV that were transmitted in a nonpersistent manner by two aphid vectors, i.e., *Aphis gossypii* and *Myzus persicae*, using 12 restriction enzyme marker bearing CMV mutants (Ali et al. 2006). The mechanically inoculated leaves used for acquisition of the virus contained all 12 of the mutants, while after transmission with either aphid species an average of three of the 12 mutants were present in the newly infected leaves. Transmission of the individual CMV mutants was largely stochastic, indicating that a bottleneck existed during the horizontal transmission of CMV. Further experiments showed that genetic variation in the artificial CMV population was not reduced during the acquisition of the virus by aphid vectors but it was significantly reduced during the inoculation period. Further work is needed with other plant viruses that are transmitted through insect vectors in a semipersistent or persistent manner.

7.4 Bottlenecks During Vertical Transmission

Approximately 20% of the reported plant viruses are transmitted from generation to generation in the seed (Mink 1993). Genetic bottlenecks may be severe in the vertically transmitted viruses because only a small number of viruses are able to cross the barriers to infect the embryo (Manrubia et al. 2005). Recently, Rebenstorf et al. (2006) hypothesized that both pollen and seed act as potential bottlenecks during the transmission of *Cherry leaf roll virus*. However, to date, there is no experimental evidence that genetic bottlenecks act to reduce viral diversity during seed transmission of plant viruses.

We are currently investigating infection of artificial populations of CMV in both zucchini and spinach plants. Our initial results indicate the presence of a narrow bottleneck during the transmission of CMV from inoculated leaves to the reproductive tissues (e.g., pollen and stigma) (Ali and Roossinck, unpublished work). However, the outcome of such studies will depend on the successful seed transmission of the virus. So far, we have obtained a very limited number of infected seeds, which makes our results inconclusive; therefore, future experimental work will clearly determine the role of genetic bottlenecks in shaping the viral population during seed transmission of plant viruses.

7.5 Genetic Drift Versus Selection

Genetic drift is the change in frequency of mutant viruses in a population that occurs only by chance effects. A major cause of population differentiation could be genetic drift as a result of genetic bottlenecks during the systemic movement or horizontal and vertical transmissions of viruses. After a bottleneck event where the population suddenly contracts to a small size, genetic drift can result in sudden changes in mutant frequency that occur independent of selection. In such instances, beneficial adaptations may be eliminated.

Genetic drift has several important effects on plant virus evolution. For example, drift reduces genetic variation in populations, which will potentially decrease a population's ability to evolve in response to new selective pressures. Genetic drift acts faster and has more drastic results in smaller populations by increasing the rates of genetic fixation and extinction. Genetic drift depends on the effective size of the population (see 7.6) and not on the census size. It has an important role in determining the frequency and fate of mutations in the effective size of the population. A study comparing strains of WSMV concluded that divergence among strains was due to both genetic drift and negative selection (Choi et al. 2001).

Genetic drift and selection are not mutually exclusive forces and usually occur concurrently. Selection is a directional process by which differential reproduction of genetically distinct mutants or variants occurs within a population and results in decreased population diversity. Selection can be positive (increase of the fittest variants in a specific environment) or negative (decrease of less fit variants). Consequently, the fit variants will have more progeny in the next generation than less fit variants. Much of the time a newly evolved mutant is of equal fitness with the progenitor, or, if altered, its fitness may be masked in the context of the quasispecies. Such a mutation is selectively neutral and its fate is determined not by selection but by chance events.

The degree to which mutants are affected by drift and selection varies according to circumstance. In a large population, where genetic drift occurs very slowly, even weak selection on a mutant may push its frequency upwards or downwards depending on whether the mutation is beneficial or harmful. However, if the population is very small, drift will predominate. In this case, weak selective effects may not be seen at all as the small changes in frequency they would produce are overshadowed by drift.

7.6 Effective Population Size

The concept of effective population size denoted by N_e was first introduced by Wright (1931). It is a basic parameter in many models of population genetics. The effective population number is a theoretical concept that indicates the size of an ideal population that will have the same genetic variance as the observed population.

Bottlenecks severely reduce N_e even though census population size (N) may recover to the former size. The effective population size is usually smaller than the census population size.

Generally, plant virus populations can be very large in the infected plant. For example, the number of TMV particles in a systemically infected tobacco leaf ranges from 10^{11} to 10^{12} (García-Arenal et al. 2003), while the estimates of founding population sizes of TMV range from one to 16 (Sacristán et al. 2003). Similarly, the estimated effective population size of *Citrus tristeza virus* isolates ranged from 32 to 263 (Sentandreu et al. 2006) and for WSMV from four to ten, which is very low compared to the estimated 10^5–10^6 virions in a single infected cell (French and Stenger 2003). This census population size is different from and less important than the effective population size because the effective population size is responsible for initiating the infection in the next generation, not the census population. Therefore, knowledge of the effective population is fundamental in understanding plant virus population structure because low effective population size increases the role of stochastic processes in dynamics of plant virus populations and evolution (French and Stenger 2003).

7.7 Muller's Ratchet in Plant Viruses

Genetic bottlenecks accelerate the gradual accumulation of deleterious mutations in the absence of recombination, an effect known as Muller's ratchet, which is an important concept in population genetics (Muller 1964). Muller predicted that when mutation rates are high and a significant proportion of mutations are deleterious, a kind of irreversible ratchet mechanism will lead to the gradual decrease in fitness of populations, leading to "mutational meltdown," when they are subjected to bottlenecks. Muller's ratchet is particularly effective in small populations of asexual organisms and for many plant viruses systemic infection and transmission leads to a bottleneck in which the resultant populations are small (Clarke et al. 1993; Li and Roossinck 2004; Ali et al. 2006). Muller's ratchet phenomenon has been shown experimentally in viral populations of the bacteriophage $\phi6$ by Chao (1990) and in a negative-sense, unsegmented animal *Vesicular stomatitis virus* (Duarte et al. 1992, 1994; Clarke et al. 1993).

There is no direct evidence of Muller's ratchet phenomenon in plant viruses. However, a study of two tobamoviruses [TMV and *Tobacco mild green mosaic virus* (TMGMV)] that infected *Nicotiana glauca* in Australia over a 100-year period could be interpreted as being due to Muller's ratchet (Fraile et al. 1997). Both viruses were analyzed from samples deposited in the New South Wales (NSW) Herbarium between 1899 and 1972, and others from living plants collected in 1985 and 1993. The authors found that many plants collected before 1950 were infected with both TMV and TMGMV. Interestingly, plants collected after that date were only infected with TMGMV. In experimental mixed infections of *N. glauca*, TMV accumulated to about 10% of the level of that in single inoculations, while the level of TMGMV was not affected. It was concluded that TMV colonized *N. glauca* in NSW earlier or faster than TMGMV. In joint infections, however, TMGMV caused a decrease of the TMV population below a threshold at which deleterious mutations were eliminated, resulting in the loss of TMV.

7.8 Bottlenecks and Speciation

Genetic bottlenecks are also responsible for founder effects, i.e., establishment of a new population by a small number of mutants, carrying only a small fraction of the original population's genetic variation. Founder effects can lead to the speciation and subsequent evolution of new species (Novella et al. 1995; Wang et al. 2006). Bottlenecks increase the rate of random genetic drift and play an active role in speciation and founding events. Thus, it is evident that the presence of a large number of strains that has been reported for various characterized plant viruses might be the results of repeated movement and transmission bottlenecks occurring frequently in nature.

7.9 Conclusions

Genetic bottlenecks have been identified in plant viral populations during systemic movement within the plant as well as horizontal transmission from plant to plant by insect vectors. Genetic drift through population bottlenecks or founder effects accompanying the spread of virus diseases may have been the main evolutionary factor in shaping population structures of plant RNA viruses. However, our understanding of the role of genetic bottlenecks in plant virus evolution is still in its early stages. Further research is required to provide complete knowledge about the role of movement and transmission bottlenecks, genetic drift, small effective population sizes, and Muller's ratchet in the speciation and evolution of plant viruses.

Acknowledgement This work was supported by the Samuel Roberts Noble Foundation.

References

Albiach-Marti MR, Mawassi M, Gowda S, Satyanarayana T, Hilf ME, Shanker S, Almira EC, Vives MC, Lopez C, Guerri J, Flores R, Moreno P, Garnsey SM, Dawson WO (2000) Sequences of *Citrus tristeza virus* separated in time and space are essentially identical. J Virol 74:6856–6865

Ali A, Li H, Schneider WL, Sherman DJ, Gray S, Smith D, Roossinck, MJ (2006) Analysis of genetic bottlenecks during horizontal transmission of *Cucumber mosaic virus*. J Virol 80(17):8345–8350

Chao L (1990) Fitness of RNA virus decreased by Muller's ratchet. Nature 348:454–455

Choi I-R, Hall JS, Henry M, Zhang L, Hein GL, French R Stenger DC (2001) Contributions of genetic drift and negative selection on the evolution of three strains of wheat streak mosaic tritimovirus. Arch Virol 146:619–628
Cillo F, Roberts IM, Palukaitis P (2002a) In situ localization and tissue distribution of the replication-associated proteins of *Cucumber mosaic virus* in tobacco and cucumber. J Virol 76(21):10654–10664
Cillo F, Roberts IM, Palukaitis P (2002b) Ultracellular localization and differential distribution of replication-associated proteins of cucumber mosaic virus in two host plant species. Paper presented at the XIIth international congress of Virology, Paris, France
Clarke D, Duarte E, Moya A, Elena S, Domingo E, Holland J (1993) Genetic bottleneck and population passages cause profound fitness differences in RNA viruses. J Virol 67:222–228
Duarte E, Clarke D, Moya A, Domingo E, Holland J (1992) Rapid fitness losses in mammalian RNA virus clones due to Muller's ratchet. Proc Natl Acad Sci USA 89:6015–6019
Duarte EA, Novella IS, Ledesma S, Clarke DK, Moya A, Elena SF, Domingo E, Holland JJ (1994) Subclonal components of consensus fitness in an RNA virus clone. J Virol 68(7):4295–4301
Fauquet CM, Mayo MA, Maniloff J, Desselberger U, Ball LA (eds) (2005) Virus taxonomy. Eighth report of the International Committee on Taxonomy of Viruses. Elsevier, San Diego
Fraile A, Escriu F, Aranda MA, Malpica JM, Gibbs AJ, García-Arenal F (1997) A century of tobamovirus evolution in an Australian population of *Nicotiana glauca*. J Virol 71(11):8316–8320
French R, Stenger DC (2003) Evolution of Wheat streak mosaic virus: dynamics of population growth within plants may explain limited variation. Annu Rev Phytopathol 41:199–214
García-Arenal F, Fraile A, Malpica J (2003) Variation and evolution of plant virus populations. Int Microbiol 6:225–232
Jerzak GVS, Bernard K, Kramer LD, Shi P-Y, Ebel GD (2007) The West Nile virus mutant spectrum is host-dependant and a determinant of mortality in mice. Virology 360(2):469–476
Li H, Roossinck MJ (2004) Genetic bottlenecks reduce population variation in an experimental RNA virus population. J Virol 78(19):10582–10587
Lucas WJ (1999) Plasmodesmata and the cell-to-cell transport of proteins and nucleoprotein complexes. J Exp Bot 50:979–987
Manrubia SC, Escarmís C, Domingo E, Lázaro E (2005) High mutation rates, bottlenecks, and robustness of RNA viral quasispecies. Gene 347:273–282
Mink GI (1993) Pollen and seed-transmitted viruses and viroids Annu Rev Phytopathol 31:375–402
Muller HJ (1964) The relation of recombination to mutational advance. Mutat Res 1:2–9
Novella IS, Duarte EA, Elena SF, Moya A, Domingo E, Holland JJ (1995) Exponential increases of RNA virus fitness during large population transmissions. Proc Natl Acad Sci USA 92:5841–5844
Novella IS, Quer J, Domingo E, Holland JJ (1999) Exponential fitness gains of RNA virus populations are limited by bottleneck effects. J Virol 73(2):1668–1671
Pirone TP, Blanc S (1996) Helper-dependent vector transmission of plant viruses. Annu Rev Phytopathol 34:227–247
Rebenstorf K, Candresse T, Dulucq MJ, Büttner C, Obermeier C (2006) Host species-dependent population structure of a pollen-borne plant virus, *Cherry leaf roll virus*. J Virol 80(5):2453–2461
Roossinck MJ (1997) Mechanisms of plant virus evolution. Annu Rev Phytopathol 35:191–209
Roossinck MJ, Schneider WL (2005) Mutant clouds and occupation of sequence space in plant RNA viruses. In: Domingo E (ed) Quasispecies: concepts and implications for virology. Current topics in microbiology and immunology, vol 299. Springer, Heidelberg, pp 337–348
Sacristán S, Malpica J, Fraile A, García-Arenal F (2003) Estimation of population bottlenecks during systemic movement of *Tobacco mosaic virus* in tobacco plants. J Virol 77(18):9906–9911
Sentandreu V, Castro JA, Ayllon MA, Rubio L, Guerri J, Gonzalez-Candelas F, Moreno P, Moya A (2006) Evolutionary analysis of genetic variation observed in citrus tristeza virus (CTV) after host passage. Arch Virol 151:875–894

Turgeon R (1996) Phloem loading and plasmodesmata. Trends Plant Sci 1(12):418–423

Vignuzzi M, Stone JK, Arnold JJ, Cameron CE, Andino R (2006) Quasispecies diversity determines pathogenesis through cooperative interactions in a viral population. Nature 439:344–348

Wang H, Huang LF, Cooper JI (2006) Analyses on mutation patterns, detection of population bottlenecks, and suggestion of deleterious-compensatory evolution among members of the genus *Potyvirus*. Arch Virol 151(8):1625–1633

Wright S (1931) Evolution in Mendalian populations. Genetics 16:97–159

Yuste E, Sanchez-Palomino S, Casado C, Domingo E, Lopez-Galindez C (1999) Drastic fitness loss in human immunodeficiency virus type 1 upon serial bottleneck events. J Virol 73:2745–2751

Chapter 8
Recombination in Plant RNA Viruses

Peter D. Nagy

Abstract Owing to increased global travel of humans carrying plants and viral vectors, introduction of new agricultural practices in combination with climate changes, the emergence of new viruses and novel viral variants is a major, continuing threat to human health and welfare. RNA recombination is one of the major forces in increasing plant virus variability and adaptation to new hosts, often leading to emergence of new variants and resistance-breaking virus strains. RNA recombination can also increase the fitness of plant RNA viruses by repairing defective viral genomes or efficiently removing deleterious mutations that result from error-prone replication. The frequency of RNA recombination is affected by several factors, including the viral replication proteins and various features of the viral RNA templates involved.

Peter Nagy
University of Kentucky, S305 201F Plant Science Building, Lexington, KY, USA
pdnagy2@uky.edu

M.J. Roossinck (ed.), *Plant Virus Evolution*.

Host genes also affect viral RNA recombination, suggesting complex interaction between a given virus and its host during viral adaptation and evolution. This chapter summarizes our current knowledge on this evolutionarily important process and its roles in emergence of new viruses or viral variants with altered pathogenecity.

8.1 Introduction

Rapid evolution of RNA viruses and the emergence of new RNA viruses, the most widespread among eukaryotic viruses, has been well documented in the last 20 years (Aaziz and Tepfer 1999; Alejska et al. 2001; Bujarski and Kaesberg 1986; Bujarski et al. 1994; Keese and Gibbs 1993; Lai 1992; Nagy and Simon 1997; Roossinck 1997; Worobey and Holmes 1999). Owing to several factors, including global travel of humans carrying plants and viral vectors, introduction of new agricultural practices and climate changes, the emergence of new viruses and novel viral variants is a major threat to human health and welfare. In addition to genetic mutations, and genome reassortment in the case of multicomponent RNA viruses, RNA recombination contributes greatly to virus genome variability. In the last 20 years since the discovery of RNA recombination in the plant virus *Brome mosaic virus* (BMV; Bujarski and Kaesberg 1986), the role of RNA recombination in RNA virus evolution has become clear. Recombination rates vary considerably among plant RNA viruses. This might be due to the different levels of precision of viral replication proteins (i.e., variations in the error-prone nature of the replicase) during RNA replication and the presence or absence of recombinationally active sequences (recombination hotspots). However, environmental and host effects likely influence the rate of RNA recombination in addition to the better characterized viral factors. Natural selection on the recombinant and parent viruses ensures the survival of only the fittest.

Recombination is the most prevalent among retroviruses, pararetroviruses, including *Cauliflower mosaic virus* (Froissart et al. 2005), and positive-stranded RNA viruses (Aaziz and Tepfer 1999; Worobey and Holmes 1999), whereas double-stranded and negative-stranded RNA viruses recombine with low frequency. This chapter will focus mostly on studies performed with plus-stranded RNA viruses, which are the most abundant among plant-infecting viruses. First, we will present the documented cases of recombination events in plant RNA viruses, then we will discuss the mechanism of RNA recombination, including the roles of the viral coded proteins, the viral RNA and host factors in RNA recombination events.

8.2 The Benefit of RNA Recombination to RNA Virus Evolution

RNA recombination is a process that joins noncontiguous segments of viral RNA(s) (Lai 1992; Nagy and Simon 1997). RNA recombination can take place between two related or unrelated RNAs or within a single RNA, the latter frequently leading to the generation of defective interfering (DI) RNAs (Hillman et al. 1987; Rochon 1991; White and Morris 1999; White and Nagy 2004). RNA recombination can promote genome rearrangement or movement of functional domains among different viruses or between the host cell and the virus, thus shaping the RNA virus world dramatically (Lai 1992; Nagy and Simon 1997; Worobey and Holmes 1999; Zimmern 1988). The benefit of RNA recombination is that new, sometimes better adapted genotypes can be formed rapidly, which can outcompete parental viral populations. RNA recombination can also increase the fitness of viruses in some hosts (Fernandez-Cuartero et al. 1994).

In contrast to its widely known role in increasing genome variability, RNA recombination functions in genome repair as well. It is thought that truncated/damaged RNA molecules could still participate in RNA recombination, leading to the repair of viral RNA molecules (Guan and Simon 2000; Hema et al. 2005; Nagy et al. 1997; Rao and Hall 1993). The repair function of RNA recombination might compensate viruses for the high mutation rate, which could introduce detrimental mutations into the viral genomes, reducing the fitness of viral populations (Roossinck 1997, 2003). Thus, RNA recombination also could be regarded as the major guardian of the viral genome, whereas its function in increasing genome variability might only be secondary, although significant.

Based on the "precision" of recombination events, RNA recombination can lead to various genetic changes. These include sequence insertions or duplications if the recombination end (break) point in one of the recombining RNAs is upstream relative to the end point on the other RNA (Fig. 8.1). Reversal of the positions of recombination end points on the viral RNAs can lead to deletions. The most well-known among viral deletion derivatives are DI RNAs, which require the parental virus for replication/spread and interfere with the accumulation of the parental virus in the host cells (White and Morris 1999). Importantly, RNA recombination can also lead to point mutations as shown schematically in Fig. 8.1. Another type of recombinant shows no sequence changes at the recombination sites when compared with the parental RNAs owing to the high precision of the recombination events. These precise recombinants can only be detected if marker mutations are introduced upstream and downstream in the parental RNAs. Because of the technical difficulties in determining these precise recombination sites, the frequency of precise recombinants in natural viral populations is unknown. It is possible that a significant fraction of viral RNA progeny might consist of these precise recombinants. Indeed, under experimental conditions, every single BMV RNA can go through precise recombination in each replication cycle (Urbanowicz et al. 2005). This and other work suggests that RNA recombination is likely a frequent event, but the detection of recombinants usually requires sophisticated approaches (Sec. 8.4).

RNA recombination is a major concern in transgenic resistance to viruses because of the theoretical possibility of the creation of new viruses formed via recombination between an infecting RNA virus and the transgene (Aaziz and Tepfer 1999; Adair and Kearney 2000; Falk and Bruening 1994; Greene and Allison 1994). This issue further emphasizes the urgent need for a better understanding of the role of RNA recombination in plant virus evolution.

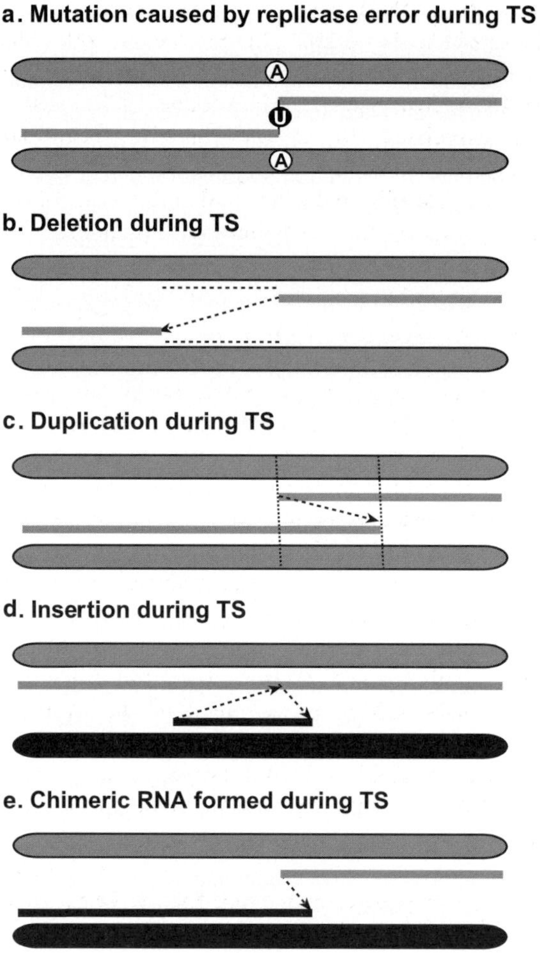

Fig. 8.1 Genetic changes caused by RNA recombination in plant RNA viruses. *TS* template switching. Gray bars represent homologous viral RNAs, whereas black bar shows heterologous RNAs

8.3 The Role of RNA Recombination in Plant Virus Variability

Naturally occurring RNA recombinants can be discovered by analyzing virus populations in natural infections and by comparison of large sequencing data sets generated with partial or complete viral genomic sequences (see Chap. 10 for appropriate methods). Accordingly, a large number of natural RNA recombinants have been described for many plant positive-stranded RNA viruses, including cucumoviruses (Bonnet et al. 2005; Chen et al. 2002) and the related ilarviruses (Ge et al. 1997), closteroviruses (Rubio et al. 2001; Vives et al. 1999, 2005), luteoviruses (Guyader and Ducray 2002; Hauser et al. 2002), nepoviruses (Le Gall et al.

1995b; Vigne et al. 2004, 2005), potyviruses (Bousalem et al. 2000; Desbiez and Lecoq 2004; Glasa and Candresse 2005; Lorenzen et al. 2006; Moreno et al. 2004; Tan et al. 2004; Zhong et al. 2005), tobamoviruses (Lartey et al. 1996) and tobraviruses (Robinson 1994; Swanson and MacFarlane 1999). Also, natural satellite RNA recombinants associated with various viruses have been found (Aranda et al. 1997; Guo et al. 2005), and natural recombinants between viruses belonging to different virus groups have also been documented on the basis of sequence comparison. A detailed analysis of the evolutionary consequences of recombination between disparate viruses is included in Chap. 9.

A phylogenetic survey of plant viruses (Chare and Holmes 2006) indicated that 12 of 36 plant RNA viruses showed evidence for recombination. In one study, 10% of the 76 isolates of *Turnip mosaic virus*, a member of the *Potyvirus* genus, were found to be recombinants (Ohshima et al. 2002). In another study with potyviruses 18 out of 109 isolates tested by Revers et al. (1996) were of recombinant origin. A significant role of RNA recombination in *Cucumber mosaic virus* (CMV) evolution was documented among natural isolates collected in Spain between 1989 and 2002 (Bonnet et al. 2005). Recombinants comprised 17% of the isolates, more than reassortants (5%), consistent with a higher biological cost of reassortment than RNA recombination. These studies indicate that recombination is relatively common in many viruses, especially in potyviruses, the largest group of plant viruses.

Virus recombinants are also easily formed in transgenic plants expressing defective or incomplete viral sequences (Aaziz and Tepfer 1999; Falk and Bruening 1994; Greene and Allison 1994). For example, inoculation of transgenic tobacco plants expressing CMV RNA 1 component with CMV RNA 2 and RNA 3 transcripts led to regeneration of CMV RNA 1 carrying 3′ noncoding sequences from either RNA 2 or RNA 3, resulting in systemic infection of the transgenic plants (Canto et al. 2001). Similarly, inoculation of transgenic *Nicotiana benthamiana* plants expressing a nontranslatable *Tobacco mosaic virus* (TMV) RNA with a defective TMV RNA gave rise to recombinant TMV RNAs in 32% of plants (Adair and Kearney 2000). Thus, viral recombination in transgenic plants can be a common event.

Yet another piece of evidence supporting frequent viral RNA recombination in plants is the existence of defective RNAs (D RNAs) or DI RNAs (reviewed in White and Morris 1999), which are intramolecular recombinants. D RNAs and DI RNAs have been found in association with an increasing number of viruses, including *Tomato bushy stunt virus* (TBSV) and other tombusviruses (Hillman et al. 1987; Rochon 1991; Rubino et al. 1990; White and Morris 1994a, b), *Turnip crinkle virus* (TCV; Li et al. 1989; Simon et al. 2004), *Broad bean mottle virus* (Pogany et al. 1995), *Bamboo mosaic virus* (Yeh et al. 1999), *Citrus tristeza virus* (Ayllon et al. 1999), *Potato yellow vein virus* (Eliasco et al. 2006), TMV (Chandrika et al. 2000; Rabindran and Dawson 2001), *Tobacco rattle virus* (Hernandez et al. 1996) and *Banana virus X* (Teycheney et al. 2005).

In summary, phylogenetics-based sequence comparisons and experimental approaches support the model that genetic recombination often causes changes in natural populations of plants viruses, resulting in enhanced pathogenecity, extended host range or overcoming host resistance factors (Escriu et al. 2003; Garcia-Arenal et al. 2003). The emerging new virus genotypes can compromise the effectiveness of antiviral strategies, underlying the importance of understanding viral RNA recombination.

8.4 Experimental Approaches to Study Viral RNA Recombination

Studying RNA recombination in experimental settings is a challenging task. This is because (1) RNA recombination, unlike genome replication, does not need to occur in every infected cell; (2) recombination is a chance event that likely occurs with variable frequencies for different viruses and strains and could be affected by the host and environmental factors; (3) the nature of recombinants and that of recombination junctions are difficult to predict or detect in the background of abundant parental (nonrecombinant) RNAs; (4) obtaining the same recombinant twice or more in independent infections can be a challenging task, so documentation and publication of recombination events requires probabilistic approaches; (5) most of the recombinants generated are likely unviable or show reduced fitness within the virus population, so detection of recombinants might be biased toward isolation of the fittest recombinants; (6) detection methods, such as reverse-transcription PCR (RT-PCR), might also contribute to generation of artifactual recombinants. To overcome the detection limit and minimize artifactual results, several sophisticated methods have been developed to study RNA recombination in experimental systems. The examples for experimentally induced recombinants include bromoviruses (Bujarski and Kaesberg 1986; Nagy and Bujarski 1992, 1993, 1995; Rao and Grantham 1994; Rao and Hall 1993), carmoviruses (Carpenter et al. 1995; Carpenter and Simon 1996; Cascone et al. 1993; Simon 1999), cucumoviruses (Masuta et al. 1998; Suzuki et al. 2003) nepoviruses (Le Gall et al. 1995a) and tombusviruses (Borja et al. 1999; Panaviene and Nagy 2003; Shapka and Nagy 2004; White and Morris 1994a, b, 1995).

8.4.1 In Vivo Approaches

RNA recombinants can be observed in whole plants or in protoplasts (single cells lacking cell walls). Because the recombinant viruses, in competition with parental viruses, have to spread cell to cell and long distance in whole plants, the generated recombinants are under considerable selection pressure in plants. A larger population of recombinants potentially could be isolated from protoplasts, where cell-to-cell movement does not take place. Yeast, a model host, has also been developed recently to study plant viral RNA recombination in a more controllable cellular environment (Garcia-Ruiz and Ahlquist 2006; Panavas and Nagy 2003).

For in vivo analysis of recombination, coinfection of the same cells with different viruses or virus strains followed by northern blotting or RT-PCR (together with cloning and sequence analyses) has been used most frequently to identify recombinants that carry partial sequences of both parent viruses. RNA recombination can also be observed between highly similar RNA molecules that carry marker mutations to allow for unambiguous identification of the recombinants. Owing to the technical difficulties, most of our understanding of RNA recombination comes from a few well-developed experimental RNA recombination systems, such as BMV, TCV, TBSV and related tombusviruses (Alejska et al. 2001; Bujarski and Nagy 1994; Nagy and Simon 1997; Simon et al. 2004; White and Nagy 2004).

8.4.2 *In Vitro Approaches*

Highly efficient in vitro RNA recombination assays have been recently developed for BMV, CMV, TCV and tombusviruses (Cheng and Nagy 2003; Cheng et al. 2005; Kim and Kao 2001; Nagy et al. 1998; Wierzchoslawski and Bujarski 2006). These include partially purified viral replicases or purified recombinant RNA-dependent RNA polymerases (RdRps). The generated recombinants were detected by denaturing polyacrylamide gel electrophoresis analysis of RdRp products, RT-PCR and sequence analysis (Cheng and Nagy 2003; Kim and Kao 2001; Wierzchoslawski and Bujarski 2006). The advantage of the in vitro approaches over the in vivo approaches is that the original recombinants can be detected, whereas in vivo the generated recombinants might go through repeated amplification, resulting in the detection of the progenies of the recombinants.

8.5 Mechanisms of RNA Recombination

The mechanisms of RNA recombination have generated significant interest in the last 15 years. The mechanistic studies could be important to predict the rate of recombination and the types of recombinants generated by various viruses. Yet, studying the mechanisms of plant RNA virus recombination turned out to be a major challenge owing to the occurrence of different classes of RNA recombinants (Sec. 8.6), the various frequencies of recombination events, removal of deleterious recombinants by natural selection pressure and the lack of unambiguous conclusions drawn from analysis of recombination junction sequences. In spite of the difficulties, however, both in vivo and in vitro evidence supports the existence of a replicase-driven template-switching (TS) mechanism. In contrast, an RNA ligation-based mechanism obtained only limited support (Alejska et al. 2001; Chetverin et al. 1997).

Evidence for the TS mechanism is abundant, including the ability of purified viral replicases or RdRps to generate RNA recombinants under defined conditions in the test tube (Cheng and Nagy 2003; Cheng et al. 2005; Kim and Kao 2001; Nagy and Simon 1998a, b; Nagy et al. 1998; Wierzchoslawski and Bujarski 2006). In addition, many of the recombination junctions characterized contain extra, nontemplated nucleotides, likely the result of addition by an error-prone viral polymerase (Cheng and Nagy 2003). Mutagenesis of viral RdRps or viral auxiliary replication proteins led to altered recombination frequencies and changed recombination sites in agreement with a TS mechanism (Figlerowicz et al. 1997, 1998; Nagy et al. 1995; Panaviene and Nagy 2003). Interestingly, mutation within the RNA-binding domain of the tombusvirus p33 replication protein affected recombination, suggesting that holding/releasing of the viral RNA template might be a key step in RNA recombination (Panaviene and Nagy 2003). Moreover, *cis*-acting replication elements, such as genomic and subgenomic promoters and replication enhancer elements, constitute recombination hotspots for many plant viruses (Cheng and Nagy 2003; Kim and Kao 2001; Nagy and Bujarski 1992; Nagy et al. 1998, 1999b; Wierzchoslawski et al. 2003), further supporting the TS model.

On the basis of mechanistic data, TS events are divided into three steps performed by the viral replicase with the involvement of the donor and acceptor RNA templates (Nagy and Simon 1997). First, the progress of complementary RNA synthesis by the viral replicase on the original RNA template, called donor RNA, might be halted owing to various factors. The pausing replicase/nascent RNA complex is thought to dissociate from the donor site and associate with an acceptor site on the same or a separate template (termed the acceptor template). Then, resumption of RNA synthesis by the viral replicase on the acceptor template will take place using the nascent RNA as a primer (Nagy and Simon 1997). The newly made recombinant RNA will then go through postrecombinational amplification if its sequence/structure is compatible with replication. Overall, the probabilistic nature of RNA recombination, including the rate of recombination and the sites of recombination breakpoints, might come from the several factors involved and their complex interactions.

Although RNA recombination occurs at high frequency in vitro (Cheng and Nagy 2003; Kim and Kao 2001), the rate of in vivo RNA recombination is likely much lower, possibly owing to the presence of one RNA or only a small number of viral RNA templates in each replicase complex (Schwartz et al. 2002). This reduced rate of RNA recombination in vivo is probably advantageous for plant viruses to minimize the potentially deleterious effects of recombination.

8.6 Classification of RNA Recombinants

8.6.1 Traditional Classification of RNA Recombinants

Traditional classification was based on the extent of identity between the recombining RNAs and on the assumption that RNA recombination is analogous to DNA recombination (Lai 1992). The RNA recombinants were defined as homologous if the two recombining RNAs were highly similar (or identical) and recombination was precise. Thus, the recombination end points were not known, only the region of recombination was known, based on the distribution of marker mutations (corresponding positions in the viral RNAs with different nucleotides). The second class, termed aberrant homologous recombinants, are similar to homologous recombinants, except the sites of recombination are imprecise, leading to minor changes, such as short deletions/insertions/mutations at the recombination sites. The third class is the nonhomologous recombinants, which resulted from recombination between two unrelated RNAs (Lai 1992).

The problem with the traditional classification is that RNA recombination is mechanistically rather different from DNA recombination. RNA recombination is based on TS events driven by the viral replicase (Sec. 8.5) and RNA recombination does not seem to utilize long stretches of identity during the recombination events (Nagy and Bujarski 1993, 1995, 1996; Nagy and Simon 1997). For example, in BMV similar lengths of sequence identity can enhance recombination if it is AU-rich (Nagy and Bujarski 1996, 1997; Shapka and Nagy 2004), but decrease recombination if it is GC-rich (Nagy and Bujarski 1998; Nagy et al. 1999a). Moreover, *cis*-acting replication elements play major roles in RNA recombination by binding to the viral replicase, making the distribution of recombination sites uneven within a particular viral genome (Nagy et al. 1998, 1999b). Analysis of natural recombinants in CMV populations also supported the suggestion that recombinants coding for hybrid proteins are at a disadvantage when compared with recombinants that exchange whole coding regions (Bonnet et al. 2005). Therefore, it is rather misleading to use the traditional classification, because homologous and aberrant homologous recombination automatically imply that sequence identity is the driving force in these types of recombination events. This frequently is not the case, but instead, similarly or uniquely positioned *cis*-acting replication elements are the major driving forces in many homologous/aberrant homologous recombination events in plant RNA viruses (Nagy and Simon 1997).

8.6.2 Nagy–Simon Classification of RNA Recombinants

More precise classification can be made by incorporating the known mechanisms leading to RNA recombination. Nagy and Simon (1997) classified RNA recombinants into three major classes (Fig. 8.2). Class 1 is the base-pairing-dependent (similarity-required) recombinants, which are generated by precise alignment between the primer RNA (the incomplete nascent strand derived from the donor template) and the acceptor RNA prior to reinitiation of primer-dependent RNA synthesis by the viral replicase (Fig. 8.2). Thus, sequence identity of 15–30 nt or longer between the donor and acceptor RNAs facilitates the generation of class 1 recombinants. Class 2 recombinants are generated by a base-pairing-independent (similarity-nonessential) mechanism (Nagy and Simon 1997). During these recombination events, *cis*-acting replication elements might facilitate rebinding of the

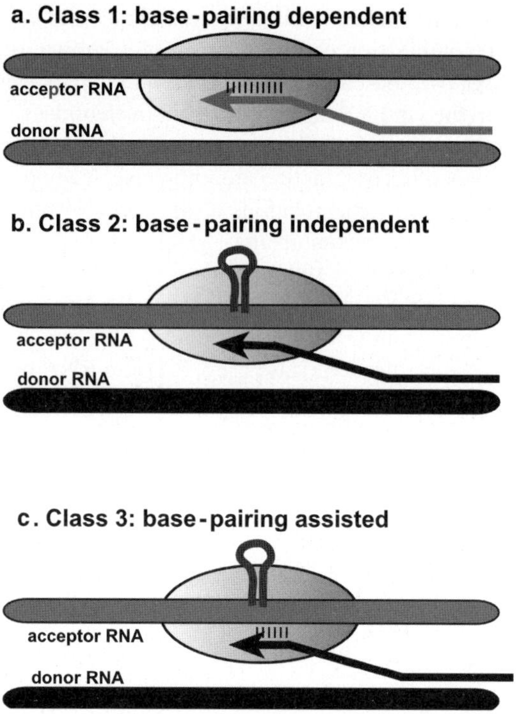

Fig. 8.2 Classification of RNA recombinants based on Nagy and Simon (1997). The TS by the viral replicase (*oval-shaped*) from the donor to the acceptor RNA is followed by primer-driven RNA synthesis (indicated by an *arrow*). Base-pairing between the primer and the acceptor template is shown. Gray bars represent homologous viral RNAs, whereas black bar shows heterologous RNAs. The *hairpin structure* symbolically represents *cis*-acting elements present on the acceptor RNA, which are proposed to facilitate binding of the viral replicase during TS events. (Modified from Flint et al. 2004)

viral replicase and primer RNA to the acceptor RNA prior to reinitiation of primer-dependent RNA synthesis by the viral replicase (Fig. 8.2). Generation of class 2 recombinants does not need sequence similarity between the donor and acceptor RNAs. The third class of recombinants, the base-pairing-assisted recombinants, are formed by a mechanism with features combining the first two classes. The *cis*-acting replication elements might facilitate rebinding of the viral replicase, but limited alignment (base-pairing) between the primer RNA and the acceptor RNA could also occur prior to reinitiation of primer-dependent RNA synthesis by the viral replicase (Fig. 8.2). Thus, short sequence similarity (3–20 nt in length) between the donor and acceptor RNA assists recombination events in class 3 recombinants (Nagy and Simon 1997). The fourth class of recombinants are theoretical: they would be formed by a completely random recombination mechanism between heterologous RNAs independent of *cis*-acting replication elements and in the absence of base-pairing. The formation of the latter class of recombinants has not been conclusively demonstrated for plant RNA viruses.

8.6.3 Occurrence of Different Classes of RNA Recombinants

Among the four classes of recombinants, the second and third classes are the most frequent for plant RNA viruses. Generation of class 1 recombinants has not yet been firmly confirmed, although recombinants that have junctions within coding regions might be formed via this mechanism. For example, in spite of the traditional use of "homologous recombination" reported in the literature, most BMV recombinants are probably not class 1 type, but class 3 type recombinants. This is because most BMV recombinants have the junction sites within noncoding sequences carrying *cis*-acting replication elements. These include the promoter regions for minus-strand initiation and for subgenomic RNA, which likely bind to the viral replicase (Nagy and Bujarski 1992, 1993; Rao and Hall 1993; Urbanowicz et al. 2005). TCV and tombusviruses also generate class 2/class 3 recombinants that are formed with the help of *cis*-acting replication elements (Nagy and Simon 1998a, b). In the case of "end-to-tail" recombinants frequently obtained in in vitro assays with BMV, TCV and tombusvirus replicases/RdRps, the RNA templates carry *cis*-acting sequences likely promote recombination formation in vitro (Cheng and Nagy 2003; Cheng et al. 2005; Kim and Kao 2001). Altogether, these observations suggest that plant RNA viruses might utilize *cis*-acting replication elements to assemble functional novel RNAs or reshuffle existing RNA genomes (Cheng et al. 2005). This approach might be more efficient as well as more flexible than recombination within coding regions, where precise alignment is absolutely essential to maintain translational reading frames and protein structure. Indeed, the replication enhancer of tombusviruses, which is a recombination hotspot, was used by the RdRp of TCV carmovirus during TS events in vitro and also for subsequent RNA synthesis. This could facilitate postrecombinational amplification of the recombinant RNAs carrying heterologous *cis*-acting sequences (Cheng et al. 2005). Thus, *cis*-acting replication/recombination elements might play central roles in modular evolution of plant RNA viruses (Fig. 8.3; see Chap. 9).

8.7 The Role of Viral Replication Proteins in RNA Recombination

The role of viral-coded replication proteins in viral RNA recombination has been studied using two approaches. First, mutations were introduced into replication proteins coded by two viruses, BMV and *Cucumber necrosis virus* (CNV), a tombusvirus closely related to TBSV. Recombination assays on *Chenopodium quinoa* leaves or in *N. benthamiana* protoplasts revealed altered rates of recombinant accumulation and a shift in the sites of recombination (Figlerowicz et al. 1997; Nagy et al. 1995; Panaviene and Nagy 2003). These data strongly support a direct role for replicase proteins in recombination events. The second assay to study the mechanism of RNA recombination in BMV, CMV, TCV and tombusviruses used purified recombinant RdRp preparations or partially purified replicases (Cheng and Nagy 2003; Cheng et al. 2005; Kim and Kao 2001; Nagy and Simon 1998a, b; Nagy

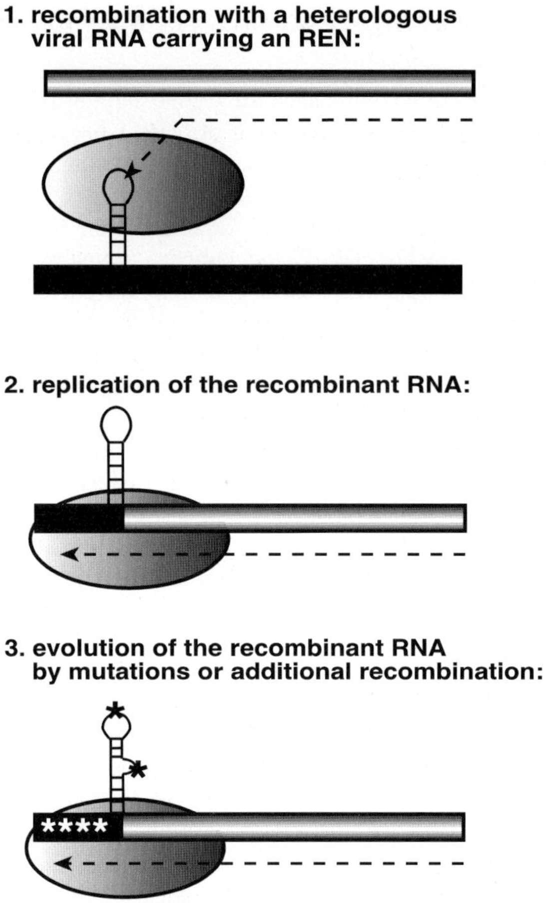

Fig. 8.3 Proposed role of heterologous *cis*-acting replication elements in promoting intervirus recombination and subsequent evolution/adaptation of the recombinant RNA virus. *REN* is a replication enhancer. (Modified from Cheng et al. 2005)

et al. 1998; Wierzchoslawski and Bujarski 2006). These replicases/RdRps generated recombinants between two different RNA templates in vitro. Interestingly, the recombinants had most of the junction sites at or near the end of the templates, albeit a few sites were internally located within the template (Cheng and Nagy 2003; Kim and Kao 2001). The distribution of recombination breakpoints suggests that the 5′ end of the viral RNA could serve as a recombination hotspot, probably by causing a strong pausing site for the replicase (which progresses from 3′ to 5′ on the template strand), thus promoting TS events. The available data on the effects of replication protein mutations on RNA recombination and the ability of the plant virus RdRps to support the generation of recombinants in vitro strongly favor the TS mechanism of RNA recombination.

8.8 Viral RNA Sequences Form Recombination Hotspots and Coldspots

8.8.1 Nonrandom Distribution of Recombination Breakpoints in the Viral RNA

In theory, each position in the viral genome should have the same probability of becoming part of a recombination junction. In contrast, experimental data show the presence of recombination hotspots and coldspots (Cheng and Nagy 2003; Nagy and Bujarski 1993, 1995, 1996, 1997; Nagy et al. 1999a; Shapka and Nagy 2004; Urbanowicz et al. 2005; White and Morris 1994b, 1995). This suggests that the sequence and/or higher-order structures of RNA likely play important roles in the selection of recombination sites (Nagy and Simon 1997). It is possible that particular RNA sequences/structures might be favored by the viral replicase during the TS events (i.e., represent hotspots). It is also possible that recombinants are generated randomly, but postrecombinational selection pressure might eliminate poorly adapted/replication-deficient recombinants, which are then only present in the initial recombination pool. The natural selection would eliminate a significant portion of the pool, thus resulting in recombinants with "artificial" hotspots that would not reflect the true distribution of recombination sites. Although selection pressure certainly affects the rates of recombinants in vivo, in vitro experiments with viral replicases/RdRps revealed the existence of authentic hotspots (Cheng and Nagy 2003; Cheng et al. 2005; Kim and Kao 2001; Nagy et al. 1999b; Wierzchoslawski and Bujarski 2006). The hotspots identified included replication enhancers and promoter sequences. These in vitro data are consistent with protoplast data that also revealed a key role for the same *cis*-acting replication elements in recombination (Cascone et al. 1993; Shapka and Nagy 2004; Urbanowicz et al. 2005). Therefore, the combination of in vitro and in vivo data supports the presence of recombination hotspots and coldspots in some plant RNA viruses.

8.8.2 RNA Determinants of RNA Recombination

It is likely that RNA determinants present in the donor and acceptor templates play different roles in RNA recombination. The donor template is expected to induce pausing/dissociation of the viral replicase, while it synthesizes the complementary RNA, which must serve as a primer in TS events. In contrast, the acceptor template is likely involved in promoting the binding of the viral replicase and/or the primer RNA (Nagy and Simon 1997).

What feature of the RNA might cause replicase pausing/dissociation on the donor template? Mechanistic studies with tombusviruses and carmoviruses revealed that 5′-ends of full-length or truncated viral RNAs frequently constitute recombination hotspots in vivo (Cheng et al. 2006; Serviene et al. 2005) and in in vitro RdRp assays (Cheng and Nagy 2003; Cheng et al. 2005) and they are also strong pausing sites for the viral replicase. The 5′ end of the template also served as a strong pausing site for the BMV replicase and constituted recombination hotspots, suggesting that the 5′ end of the template might serve as a general recombination hotspot (Kim and Kao 2001). The truncated RNA templates could be generated by limited RNA degradation by a cellular RNase (Cheng et al. 2006) or internal initiation of the viral replicase on full-length templates (Panavas et al. 2002), which would lead to production of truncated RNA templates in subsequent rounds of replication.

Another RNA determinant that might promote pausing/dissociation of the viral replicase on the donor template is sequence composition. For instance, during strand elongation, weak A–U base-pairing between the template and the nascent (primer) strand within an AU stretch might promote replicase pausing/dissociation, as suggested for BMV (Nagy and Bujarski 1996, 1997) and HIV (DeStefano et al. 1994; Wu et al. 1995). Also, tombusvirus replicase efficiently supported TS within an AU-rich sequence in the in vitro RdRp assay (Cheng and Nagy 2003). The same AU-rich sequence was found to promote tombusviral RNA recombination in protoplasts and BMV recombination in plants (Nagy and Bujarski 1997; Shapka and Nagy 2004). Sequence composition in the donor template may therefore represent a general determinant for recombination in several viral systems.

The third type of RNA element that could facilitate replicase pausing/dissociation and RNA recombination is stable RNA structure. Difficulty in unwinding of stable RNA structures by the viral replicase might promote replicase pausing and TS events (Nagy and Bujarski 1993), as demonstrated for various plant viruses (Havelda et al. 1997; White and Morris 1995). Formation of stable RNA structures between two different RNAs (termed heteroduplex formation) could also facilitate bringing and keeping the donor and acceptor RNAs in close vicinity prior to TS events (Nagy and Bujarski 1993).

In addition to the abovementioned general RNA features, it is likely that other, less universal elements might play roles in recombination when present in the donor template. For example, replication enhancers in tombusviruses and carmoviruses are known to affect recombination when present on the donor template. However, it is still unknown how these sequences could facilitate the participation of the donor template in recombination.

What feature of the acceptor RNA might promote the binding of the viral replicase and/or the primer RNA? The RNA determinants in the acceptor RNA that influence recombination have been characterized using in vitro assays based on CNV and TCV RdRps. Protein-RNA binding assays demonstrated that replication enhancers, which serve as recombination hotspots, are superior in binding to the viral replicase proteins when compared with other viral or nonviral sequences of similar length (Nagy et al. 1999b; Panavas and Nagy 2005; Panavas et al. 2003). Therefore, replication enhancers and promoters might constitute recombination hotspots because of their high affinity for the viral replicase proteins. An RNA element termed internal replication element (IRE) that promotes recombination in tombusviruses (Cheng et al. 2006) also shows high-affinity binding to the replicase proteins (Nagy and Pogany 2006; Pogany et al. 2005). The same IRE serves as the *cis*-acting element for template recruitment by selectively binding to the tombusvirus replication protein (Monkewich et al. 2005; Pogany et al. 2005). Interestingly, an intercistronic sequence in BMV RNA 3, containing a template recruitment element and a subgenomic promoter element, also acts as a recombination hotspot (Wierzchoslawski and Bujarski 2006; Wierzchoslawski et al. 2003), suggesting a similar role to the tombusvirus IRE. Efficient binding of the replicase to the acceptor RNA could help in the landing of the dissociated replicase/primer RNA complex, and thus promote RNA recombination.

Another feature helping acceptor–replicase association could be base-pair formation between the acceptor RNA and the primer RNA in a region of local sequence complementarity. Unfortunately, the effect of the length of sequence complementarity on RNA recombination is currently poorly defined. It is known that the BMV, TCV and CNV RdRps favor primers with base-paired regions of 4–8 bp in length, with shorter or longer primers being less effective (Cheng et al. 2002; Kim and Kao 2001; Nagy and Simon 1998a, b). These data suggest that formation of short priming regions between the acceptor sequence and the 3′ end of the primer RNA could facilitate recombination *in vivo*. Accordingly, the *in vivo* isolated TCV and tombusvirus recombinants frequently contained 1–5-nt-long sequence identity at the recombination sites, which might be the "fingerprint" of the priming events. BMV replicase might readily accept longer primer/acceptor regions (Wierzchoslawski and Bujarski 2006), which could increase the precision of RNA recombination. Accordingly, BMV supports homologous recombination with higher efficiency than TBSV or TCV (Nagy and Bujarski 1995, 1996; Urbanowicz et al. 2005). It is not yet known if the BMV homologous recombinants are formed via a base-pairing-dependent (similarity-essential) or a base-pairing-assisted recombination mechanism.

Overall, the published data suggest that the nature of the acceptor template plays a major role during landing of the dissociated replicase/primer RNA complex. The factors involved are (1) limited base-pairing between the acceptor region and the primer RNA and (2) *cis*-acting replication elements, such as replication enhancers, IRE and genomic and subgenomic promoters in tombusviruses (Cheng and Nagy 2003; Cheng et al. 2005, 2006), TCV (Nagy et al. 1998, 1999b), *Barley yellow dwarf virus* (Miller et al. 1995) and BMV (Wierzchoslawski et al. 2003) that might facilitate binding of the dissociated replicase/primer RNA complex to the acceptor region. In spite of the detailed studies on features of donor and acceptor RNAs, we still do not know if the primer RNA (recombination intermediate) is associated with the viral replicase or is free during the TS events. Further studies are needed to dissect the detailed roles of RNA intermediates and proteins in viral RNA recombination and to expand our knowledge for more plant RNA viruses that will help define the unique as well as general features/elements of RNA recombination.

8.9 The Role of the Host Genes in RNA Recombination

Several indirect observations suggested the role of the host organism in viral RNA recombination. For example, detection of TBSV RNA recombinants was easier in *N. benthamiana* plants than in pepper plants, suggesting that the generation of viral

RNA recombinants might be host-dependent (Desvoyes and Scholthof 2002). *Alstroemeria* plants supported specific recombinant CMV, whereas tobacco plants did not, suggesting that recombination might facilitate the adaptation of CMV to *Alstroemeria* plants (Chen et al. 2002). Specific antiviral defenses in the host plants might also favor the generation/accumulation of new recombinants that could elude host defenses. Infections of transgenic *N. benthamiana* plants expressing the helper-component proteinase (HC-Pro) gene from *Plum pox virus* (PPV) (which, owing to RNA silencing, were resistant to PPV) with a chimeric *Potato virus X* (PVX) carrying the PPV HC-Pro rapidly led to the generation of PVX recombinants (Barajas et al. 2006). The recombinant PVX lacked HC-Pro sequences and the recombinants were able to infect the resistant *N. benthamiana* plants systemically. This work demonstrated that recombination could help plant RNA viruses to escape host resistance.

8.9.1 Identification of Host Genes Involved in RNA Recombination Based on Yeast

To test how individual host genes could affect RNA recombination, a TBSV replicon-based recombination system was used in a model yeast host. Systematic testing of 80% of yeast genes using a single gene-deletion library revealed that TBSV recombination was increased approximately 10–50-fold in the absence of five yeast genes (Serviene et al. 2005). In contrast, deletion of four genes inhibited RNA recombination, whereas deletion of two other genes changed the profile (i.e., the sizes) of the recombinants (Serviene et al. 2005). Testing the effect of 800 of the 1,100 known essential genes of yeast (those that cannot be deleted without losing yeast viability) after their downregulation from a titratable promoter revealed that 16 host genes affected the accumulation of recombinants (Serviene et al. 2006). Overall, 95% of yeast genes were tested, and 32 of the approximately 5,500 yeast genes (0.5%) affected RNA recombination. This is a surprisingly large number of genes, suggesting that the host plays a more complex role in viral recombination than previously thought. Moreover, the systematic screen based on gene deletion/downregulation likely underestimated the number of host genes affecting RNA recombination, because (1) the functions of many host genes are complemented by those of other related genes (e.g., multimember gene families), thus the effect of single gene deletion is masked; and (2) the TBSV replicon-based screens were targeted to identify "nonhomologous" recombinants, while the generation of precise homologous recombinants was not tested (Serviene et al. 2006). The host genes identified affected the ratio of recombinant to parental RNAs; thus, these factors are not host factors affecting TBSV replication per se, since the ratio of recombinant to parental RNAs would not change if the host factor decreased replication in general. It is possible that the host genes affect the frequency of RNA recombination and/or the selection of the recombinants over the parental RNAs. The possible increased fitness of the recombinants in the given yeast strain would be interesting because the TBSV replicon was a high replication competent DI RNA (Serviene et al. 2005,

2006). Therefore, the yeast experiments suggest that TBSV RNA might need adaptation if replicating in a particular genetically modified host.

8.9.2 Function of Host Genes in RNA Recombination

The functions of the identified yeast genes in viral RNA recombination have only been studied in some detail for one gene (Cheng et al. 2006). The selected gene, *XRN1*, codes for a 5′–3′ exoribonuclease involved in messenger RNA, small interfering RNA and microRNA degradation. Deletion of *XRN1*, led to approximately 50-fold increased recombinant RNA accumulation in yeast (Serviene et al. 2005). The generated recombinants were chimeric RNAs with overlapping sequences (Fig. 8.4). Interestingly, partially degraded TBSV RNAs were detected in *xrn1Δ* yeast, which could serve as recombination intermediates (Fig. 8.4). Another interesting feature of these partially degraded viral RNAs is that they have IRE, a known recombination hotspot, "exposed" at their 5′ ends (Shapka and Nagy 2004). The IRE recombination hotspot sequence is involved in binding to the tombusvirus p33 replication protein (Pogany et al. 2005), likely facilitating recombination by increasing the local concentration of the viral replicase that contains the p33 protein, in addition to $p92^{pol}$ and host proteins (Serva and Nagy 2006). The partially degraded RNAs are likely generated via a cleavage(s) by endoribonucleases, such as Ngl2p (Cheng et al. 2006). However, the partially degraded viral RNAs could be rapidly degraded by Xrn1p in wild-type yeast, thus reducing (i.e., suppressing) the chance for partially degraded viral RNAs to participate in RNA recombination. In contrast, partially degraded viral RNAs with the recombination hotspot region exposed at the 5′ end are abundant in *xrn1Δ* yeast, apparently enhancing the frequency of recombination events. On the basis of these data, Xrn1p is a strong suppressor of RNA recombination in wild-type yeast.

Altogether, it is likely that Xrn1p affects RNA recombination by regulating the number of recombination intermediates available for recombination events in wild-type cells. It is possible that other host factors identified also affect the amounts of RNA substrates available for recombination or modify the activity of the viral replicase, thus affecting the precision of RNA synthesis and the frequency of TS. It is also possible that some host proteins might affect the fitness of the generated recombinants compared with that of the parental RNAs, thus indirectly affecting RNA recombination. Regardless of the mechanism, the complex role of host proteins in viral RNA recombination promotes the idea that viral RNA recombination is a major mechanism for viruses to adapt to new hosts.

8.10 Conclusions

RNA recombination is one of the major forces in increasing plant virus variability and adaptation to new hosts, often leading to emergence of new variants and resistance-breaking virus strains. RNA recombination can also increase viral fitness by repairing

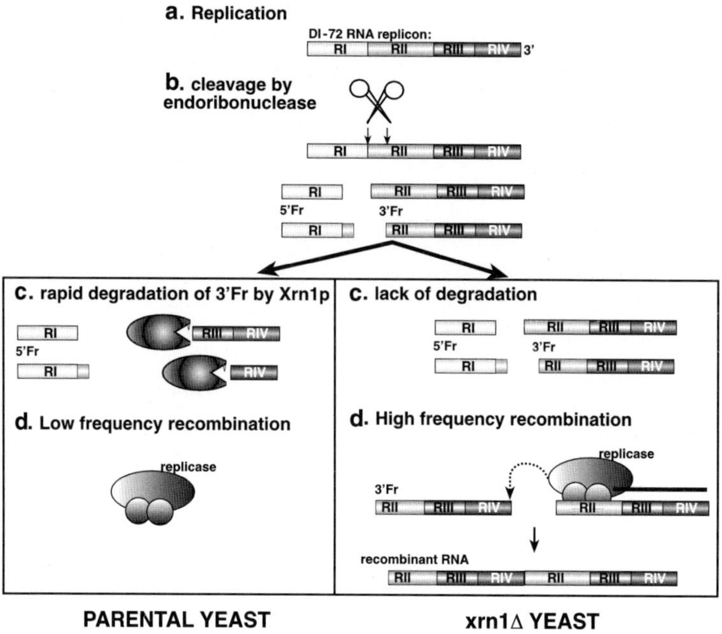

Fig. 8.4 A model for suppression of tombusviral RNA recombination by Xrn1p exoribonuclease. The partially degraded viral RNAs (generated via endoribonuclease cleavage, indicated by *scissors*) are rapidly degraded in wild-type cells, whereas they are more stable in cells lacking Xrn1p (*right panel*). Owing to the presence of cis-acting replication/recombination element (termed *RII*), the partially degraded viral RNAs then serve as recombination substrates to generate recombinant RNAs. (Modified from Cheng et al. 2006)

defective viral genomes or efficiently removing deleterious mutations that result from error-prone replication. RNA recombination is affected by several factors, including the viral replication proteins and various features of the viral RNA templates involved. Host genes also affect RNA recombination, suggesting complex interaction between a given virus and its host during viral adaptation and evolution. Future research will likely unravel further details of this evolutionarily important process in emergence of new viruses or viral variants with altered pathogenecity.

References

Aaziz R, Tepfer M (1999) Recombination in RNA viruses and in virus-resistant transgenic plants. J Gen Virol 80(6):1339–1346

Adair TL, Kearney CM (2000) Recombination between a 3-kilobase tobacco mosaic virus transgene and a homologous viral construct in the restoration of viral and nonviral genes. Arch Virol 145(9):1867–1883

Alejska M, Kurzyniska-Kokorniak A, Broda M, Kierzek R, Figlerowicz M (2001) How RNA viruses exchange their genetic material. Acta Biochim Pol 48(2):391–407

Aranda MA, Fraile A, Dopazo J, Malpica JM, Garcia-Arenal F (1997) Contribution of mutation and RNA recombination to the evolution of a plant pathogenic RNA. J Mol Evol 44(1):81–88

Ayllon MA, Lopez C, Navas-Castillo J, Mawassi M, Dawson WO, Guerri J, Flores R, Moreno P (1999) New defective RNAs from citrus tristeza virus: evidence for a replicase-driven template switching mechanism in their generation. J Gen Virol 80(3):817–821

Barajas D, Tenllado F, Diaz-Ruiz JR (2006) Characterization of the recombinant forms arising from a Potato virus X chimeric virus infection under RNA silencing pressure. Mol Plant Microbe Interact 19(8):904–913

Bonnet J, Fraile A, Sacristan S, Malpica JM, Garcia-Arenal F (2005) Role of recombination in the evolution of natural populations of Cucumber mosaic virus, a tripartite RNA plant virus. Virology 332(1):359–368

Borja M, Rubio T, Scholthof HB, Jackson AO (1999) Restoration of wild-type virus by double recombination of tombusvirus mutants with a host transgene. Mol Plant Microbe Interact 12(2):153–162

Bousalem M, Douzery EJ, Fargette D (2000) High genetic diversity, distant phylogenetic relationships and intraspecies recombination events among natural populations of Yam mosaic virus: a contribution to understanding potyvirus evolution. J Gen Virol 81(1):243–255

Bujarski JJ, Kaesberg P (1986) Genetic recombination between RNA components of a multipartite plant virus. Nature 321(6069):528–531

Bujarski JJ, Nagy PD (1994) Targeting of the site of nonhomologous genetic recombination in brome mosaic virus. Arch Virol Suppl 9:231–238

Bujarski JJ, Nagy PD, Flasinski S (1994) Molecular studies of genetic RNA-RNA recombination in brome mosaic virus. Adv Virus Res 43:275–302

Canto T, Choi SK, Palukaitis P (2001) A subpopulation of RNA 1 of Cucumber mosaic virus contains 3′ termini originating from RNAs 2 or 3. J Gen Virol 82(4):941–945

Carpenter CD, Simon AE (1996) In vivo restoration of biologically active 3′ ends of virus-associated RNAs by nonhomologous RNA recombination and replacement of a terminal motif. J Virol 70(1);478–486

Carpenter CD, Oh JW, Zhang C, Simon AE (1995) Involvement of a stem-loop structure in the location of junction sites in viral RNA recombination. J Mol Biol 245(5):608–622

Cascone PJ, Haydar TF, Simon AE (1993) Sequences and structures required for recombination between virus-associated RNAs. Science 260(5109):801–805

Chandrika R, Rabindran S, Lewandowski DJ, Manjunath KL, Dawson WO (2000) Full-length tobacco mosaic virus RNAs and defective RNAs have different 3′ replication signals. Virology 273(1):198–209

Chare ER, Holmes EC (2006) A phylogenetic survey of recombination frequency in plant RNA viruses. Arch Virol 151(5):933–946

Chen YK, Goldbach R, Prins M (2002). Inter- and intramolecular recombinations in the cucumber mosaic virus genome related to adaptation to alstroemeria. J Virol 76(8):4119–4124

Cheng CP, Nagy PD (2003) Mechanism of RNA recombination in carmo- and tombusviruses: evidence for template switching by the RNA-dependent RNA polymerase in vitro. J Virol 77(22):12033–12047

Cheng CP, Pogany J, Nagy PD (2002) Mechanism of DI RNA formation in tombusviruses: dissecting the requirement for primer extension by the tombusvirus RNA dependent RNA polymerase in vitro. Virology 304(2):460–473

Cheng CP, Panavas T, Luo G, Nagy PD (2005) Heterologous RNA replication enhancer stimulates in vitro RNA synthesis and template-switching by the carmovirus, but not by the tombusvirus, RNA-dependent RNA polymerase: implication for modular evolution of RNA viruses. Virology 341(1):107–121

Cheng CP, Serviene E, Nagy PD (2006) Suppression of viral RNA recombination by a host exoribonuclease. J Virol 80(6):2631–2640

Chetverin AB, Chetverina HV, Demidenko AA, Ugarov VI (1997) Nonhomologous RNA recombination in a cell-free system: evidence for a transesterification mechanism guided by secondary structure. Cell 88(4):503–513

Desbiez C, Lecoq H (2004) The nucleotide sequence of Watermelon mosaic virus (WMV, Potyvirus) reveals interspecific recombination between two related potyviruses in the 5′ part of the genome. Arch Virol 149(8):1619–1632

DeStefano JJ, Bambara RA, Fay PJ (1994) The mechanism of human immunodeficiency virus reverse transcriptase-catalyzed strand transfer from internal regions of heteropolymeric RNA templates. J Biol Chem 269(1):161–168

Desvoyes B, Scholthof HB (2002) Host-dependent recombination of a Tomato bushy stunt virus coat protein mutant yields truncated capsid subunits that form virus-like complexes which benefit systemic spread. Virology 304(2):434–442

Eliasco E, Livieratos IC, Muller G, Guzman M, Salazar LF, Coutts RH (2006) Sequences of defective RNAs associated with potato yellow vein virus. Arch Virol 151(1):201–204

Escriu F, Fraile A, Garcia-Arenal F (2003) The evolution of virulence in a plant virus. Evol Int J Org Evol 57(4):755–765

Falk BW, Bruening G (1994). Will transgenic crops generate new viruses and new diseases? Science 263(5152):1395–1396

Fernandez-Cuartero B, Burgyan J, Aranda MA, Salanki K, Moriones E, Garcia-Arenal F (1994) Increase in the relative fitness of a plant virus RNA associated with its recombinant nature. Virology 203(2):373–377

Figlerowicz M, Nagy PD, Bujarski JJ (1997) A mutation in the putative RNA polymerase gene inhibits nonhomologous, but not homologous, genetic recombination in an RNA virus. Proc Natl Acad Sci USA 94(5):2073–2078

Figlerowicz M, Nagy PD, Tang N, Kao CC, Bujarski JJ (1998) Mutations in the N terminus of the brome mosaic virus polymerase affect genetic RNA-RNA recombination. J Virol 72(11):9192–9200

Flint SJ, Enquist LW, Racaniello VR, Skalka AM (2004) Principles of virology. American Society for Microbiology, Washington

Froissart R, Roze D, Uzest M, Galibert L, Blanc S, Michalakis Y (2005) Recombination every day: abundant recombination in a virus during a single multi-cellular host infection. PLoS Biol 3(3):e89

Garcia-Arenal F, Fraile A, Malpica JM (2003) Variation and evolution of plant virus populations. Int Microbiol 6(4):225–232

Garcia-Ruiz H, Ahlquist P (2006) Inducible yeast system for Viral RNA recombination reveals requirement for an RNA replication signal on both parental RNAs. J Virol 80(17):8316–8328

Ge X, Scott SW, Zimmerman MT (1997) The complete sequence of the genomic RNAs of spinach latent virus. Arch Virol 142(6):1213–1226

Glasa M, Candresse T (2005) Partial sequence analysis of an atypical Turkish isolate provides further information on the evolutionary history of Plum pox virus (PPV). Virus Res 108(1–2):199–206

Greene AE, Allison RF (1994) Recombination between viral RNA and transgenic plant transcripts. Science 263(5152):1423–1425

Guan H, Simon AE (2000) Polymerization of nontemplate bases before transcription initiation at the 3′ ends of templates by an RNA-dependent RNA polymerase: an activity involved in 3′ end repair of viral RNAs. Proc Natl Acad Sci USA 97(23):12451–12456

Guo LH, Cao YH, Li DW, Niu SN, Cai ZN, Han CG, Zhai YF, Yu JL (2005) Analysis of nucleotide sequences and multimeric forms of a novel satellite RNA associated with beet black scorch virus. J Virol 79(6):3664–3674

Guyader S, Ducray DG (2002) Sequence analysis of Potato leafroll virus isolates reveals genetic stability, major evolutionary events and differential selection pressure between overlapping reading frame products. J Gen Virol 83(7):1799–1807

Hauser S, Stevens M, Beuve M, Lemaire O (2002) Biological properties and molecular characterization of beet chlorosis virus (BChV). Arch Virol 147(4):745–762

Havelda Z, Dalmay T, Burgyan J (1997) Secondary structure-dependent evolution of Cymbidium ringspot virus defective interfering RNA. J Gen Virol 78(6):1227–1234

Hema M, Gopinath K, Kao C (2005) Repair of the tRNA-like CCA sequence in a multipartite positive-strand RNA virus. J Virol 79(3):1417–1427

Hernandez C, Carette JE, Brown DJ, Bol JF (1996) Serial passage of tobacco rattle virus under different selection conditions results in deletion of structural and nonstructural genes in RNA 2. J Virol 70(8):4933–4940

Hillman BI, Carrington JC, Morris J (1987) A defective interfering RNA that contains a mosaic of a plant virus genome. Cell 51(3):427–433

Keese P, Gibbs A (1993) Plant viruses: master explorers of evolutionary space. Curr Opin Genet Dev 3(6):873–877

Kim MJ, Kao C (2001) Factors regulating template switch in vitro by viral RNA-dependent RNA polymerases: implications for RNA-RNA recombination. Proc Natl Acad Sci USA 98(9):4972–4977

Lai MM (1992) RNA recombination in animal and plant viruses. Microbiol Rev 56(1):61–79

Lartey RT, Voss TC, Melcher U (1996) Tobamovirus evolution: gene overlaps, recombination, and taxonomic implications. Mol Biol Evol 13(10):1327–1338

Le Gall O, Candresse T, Dunez J (1995a) Transfer of the 3′ non-translated region of grapevine chrome mosaic virus RNA-1 by recombination to tomato black ring virus RNA-2 in pseudorecombinant isolates. J Gen Virol 76(5):1285–1289

Le Gall OL, Lanneau M, Candresse T, Dunez J (1995b) The nucleotide sequence of the RNA-2 of an isolate of the English serotype of tomato black ring virus: RNA recombination in the history of nepoviruses. J Gen Virol 76(5):1279–1283

Li XH, Heaton LA, Morris TJ, Simon AE (1989) Turnip crinkle virus defective interfering RNAs intensify viral symptoms and are generated de novo. Proc Natl Acad Sci USA 86(23):9173–9177

Lorenzen JH, Meacham T, Berger PH, Shiel PJ, Crosslin JM, Hamm PB, Kopp H (2006) Whole genome characterization of Potato virus Y isolates collected in the western USA and their comparison to isolates from Europe and Canada. Arch Virol 151(6):1055–1074

Masuta C, Ueda S, Suzuki M, Uyeda I (1998) Evolution of a quadripartite hybrid virus by interspecific exchange and recombination between replicase components of two related tripartite RNA viruses. Proc Natl Acad Sci USA 95(18):10487–10492

Miller WA, Dinesh-Kumar SP, Paul CP (1995) Luteovirus gene expression. Crit Rev Plant Sci 14:179–211

Monkewich S, Lin HX, Fabian MR, Xu W, Na H, Ray D, Chernysheva OA, Nagy PD, White KA (2005) The p92 polymerase coding region contains an internal RNA element required at an early step in Tombusvirus genome replication. J Virol 79(8):4848–4858

Moreno IM, Malpica JM, Diaz-Pendon JA, Moriones E, Fraile A, Garcia-Arenal F (2004) Variability and genetic structure of the population of watermelon mosaic virus infecting melon in Spain. Virology 318(1):451–460

Nagy PD, Bujarski JJ (1992) Genetic recombination in brome mosaic virus: effect of sequence and replication of RNA on accumulation of recombinants. J Virol 66(11):6824–6828

Nagy PD, Bujarski JJ (1993) Targeting the site of RNA-RNA recombination in brome mosaic virus with antisense sequences. Proc Natl Acad Sci USA 90(14):6390–6394

Nagy PD, Bujarski JJ (1995) Efficient system of homologous RNA recombination in brome mosaic virus: sequence and structure requirements and accuracy of crossovers. J Virol 69(1):131–140

Nagy PD, Bujarski JJ (1996) Homologous RNA recombination in brome mosaic virus: AU-rich sequences decrease the accuracy of crossovers. J Virol 70(1):415–426

Nagy PD, Bujarski JJ (1997) Engineering of homologous recombination hotspots with AU-rich sequences in brome mosaic virus. J Virol 71(5):3799–3810

Nagy PD, Bujarski JJ (1998) Silencing homologous RNA recombination hot spots with GC-rich sequences in brome mosaic virus. J Virol 72(2):1122–1130

Nagy PD, Pogany J (2006) Yeast as a model host to dissect functions of viral and host factors in tombusvirus replication. Virology 344(1):211–220

Nagy PD, Simon AE (1997) New insights into the mechanisms of RNA recombination. Virology 235(1):1–9

Nagy PD, Simon AE (1998a) In vitro characterization of late steps of RNA recombination in turnip crinkle virus. I. Role of motif1-hairpin structure. Virology 249(2):379–392

Nagy PD, Simon AE (1998b) In vitro characterization of late steps of RNA recombination in turnip crinkle virus. II. The role of the priming stem and flanking sequences. Virology 249(2):393–405

Nagy PD, Dzianott A, Ahlquist P, Bujarski JJ (1995) Mutations in the helicase-like domain of protein 1a alter the sites of RNA-RNA recombination in brome mosaic virus. J Virol 69(4):2547–2556

Nagy PD, Carpenter CD, Simon AE (1997) A novel 3′-end repair mechanism in an RNA virus. Proc Natl Acad Sci USA 94(4):1113–1118

Nagy PD, Zhang C, Simon AE (1998) Dissecting RNA recombination in vitro: role of RNA sequences and the viral replicase. EMBO J 17(8):2392–2403

Nagy PD, Ogiela C, Bujarski JJ (1999a) Mapping sequences active in homologous RNA recombination in brome mosaic virus: prediction of recombination hot spots. Virology 254(1):92–104

Nagy PD, Pogany J, Simon AE (1999b) RNA elements required for RNA recombination function as replication enhancers in vitro and in vivo in a plus-strand RNA virus. EMBO J 18(20):5653–5665

Ohshima K, Yamaguchi Y, Hirota R, Hamamoto T, Tomimura K, Tan Z, Sano T, Azuhata F, Walsh JA, Fletcher J, Chen J, Gera A, Gibbs A (2002) Molecular evolution of Turnip mosaic virus: evidence of host adaptation, genetic recombination and geographical spread. J Gen Virol 83(6):1511–1521

Panavas T, Nagy PD (2003) Yeast as a model host to study replication and recombination of defective interfering RNA of Tomato bushy stunt virus. Virology 314(1):315–325

Panavas T, Nagy PD (2005) Mechanism of stimulation of plus-strand synthesis by an RNA replication enhancer in a tombusvirus. J Virol 79(15):9777–9785

Panavas T, Pogany J, Nagy PD (2002) Internal initiation by the cucumber necrosis virus RNA-dependent RNA polymerase is facilitated by promoter-like sequences. Virology 296(2):275–287

Panavas T, Panaviene Z, Pogany J, Nagy PD (2003) Enhancement of RNA synthesis by promoter duplication in tombusviruses. Virology 310(1):118–129

Panaviene Z, Nagy PD (2003) Mutations in the RNA-binding domains of tombusvirus replicase proteins affect RNA recombination in vivo. Virology 317(2):359–372

Pogany J, Romero J, Huang Q, Sgro JY, Shang H, Bujarski JJ (1995) De novo generation of defective interfering-like RNAs in broad bean mottle bromovirus. Virology 212(2):574–586

Pogany J, White KA, Nagy PD (2005) Specific binding of tombusvirus replication protein p33 to an internal replication element in the viral RNA is essential for replication. J Virol 79(8):4859–4869

Rabindran S, Dawson WO (2001) Assessment of recombinants that arise from the use of a TMV-based transient expression vector. Virology 284(2):182–189

Rao AL, Grantham GL (1994) Amplification in vivo of brome mosaic virus RNAs bearing 3′ noncoding region from cucumber mosaic virus. Virology 204(1):478–481

Rao AL, Hall TC (1993) Recombination and polymerase error facilitate restoration of infectivity in brome mosaic virus. J Virol 67(2):969–979

Revers F, Le Gall O, Candresse T, Le Romancer M, Dunez J (1996) Frequent occurrence of recombinant potyvirus isolates. J Gen Virol 77(8):1953–1965

Robinson DJ (1994) Sequences at the ends of RNA-2 of I6, a recombinant tobravirus Arch Virol Suppl 9:245–51

Rochon DM (1991) Rapid de novo generation of defective interfering RNA by cucumber necrosis virus mutants that do not express the 20-kDa nonstructural protein. Proc Natl Acad Sci USA 88(24):11153–11157

Roossinck MJ (1997) Mechanisms of plant virus evolution. Annu Rev Phytopathol 35:191–209

Roossinck MJ (2003) Plant RNA virus evolution. Curr Opin Microbiol 6(4):406–409

Rubino L, Burgyan J, Grieco F, Russo M (1990) Sequence analysis of cymbidium ringspot virus satellite and defective interfering RNAs. J Gen Virol 71(8):1655–1660

Rubio L, Ayllon MA, Kong P, Fernandez A, Polek M, Guerri J, Moreno P, Falk BW (2001) Genetic variation of Citrus tristeza virus isolates from California and Spain: evidence for mixed infections and recombination. J Virol 75(17):8054–8062

Schwartz M, Chen J, Janda M, Sullivan M, den Boon J, Ahlquist P (2002) A positive-strand RNA virus replication complex parallels form and function of retrovirus capsids. Mol Cell 9(3):505–514

Serva S, Nagy PD (2006) Proteomics analysis of the tombusvirus replicase: Hsp70 molecular chaperone is associated with the replicase and enhances viral RNA replication. J Virol 80(5):2162–2169

Serviene E, Shapka N, Cheng CP, Panavas T, Phuangrat B, Baker J, Nagy PD (2005) Genome-wide screen identifies host genes affecting viral RNA recombination. Proc Natl Acad Sci USA 102(30):10545–10550

Serviene E, Jiang Y, Cheng CP, Baker J, Nagy PD (2006) Screening of the yeast yTHC collection identifies essential host factors affecting tombusvirus RNA recombination. J Virol 80(3):1231–1241

Shapka N, Nagy PD (2004) The AU-rich RNA recombination hot spot sequence of Brome mosaic virus is functional in tombusviruses: implications for the mechanism of RNA recombination. J Virol 78(5):2288–2300

Simon AE (1999) Replication, recombination, and symptom-modulation properties of the satellite RNAs of turnip crinkle virus. Curr Top Microbiol Immunol 239:19–36

Simon AE, Roossinck MJ, Havelda Z (2004) Plant virus satellite and defective interfering RNAs: new paradigms for a new century. Annu Rev Phytopathol 42:415–437

Suzuki M, Hibi T, Masuta C (2003) RNA recombination between cucumoviruses: possible role of predicted stem-loop structures and an internal subgenomic promoter-like motif. Virology 306(1):77–86

Swanson MM, MacFarlane SA (1999) The E116 isolate of Dutch pea early-browning virus is a recombinant virus. Virus Res 60(1):87–94

Tan Z, Wada Y, Chen J, Ohshima K (2004) Inter- and intralineage recombinants are common in natural populations of Turnip mosaic virus. J Gen Virol 85(9):2683–2696

Teycheney PY, Marais A, Svanella-Dumas L, Dulucq MJ, Candresse T (2005) Molecular characterization of banana virus X (BVX), a novel member of the Flexiviridae family. Arch Virol 150(9):1715–1727

Urbanowicz A, Alejska M, Formanowicz P, Blazewicz J, Figlerowicz M, Bujarski JJ (2005) Homologous crossovers among molecules of brome mosaic bromovirus RNA1 or RNA2 segments in vivo. J Virol 79(9):5732–5742

Vigne E, Bergdoll M, Guyader S, Fuchs M (2004) Population structure and genetic variability within isolates of Grapevine fanleaf virus from a naturally infected vineyard in France: evidence for mixed infection and recombination. J Gen Virol 85(8):2435–2445

Vigne E, Demangeat G, Komar V, Fuchs M (2005) Characterization of a naturally occurring recombinant isolate of Grapevine fanleaf virus. Arch Virol 150(11):2241–2255

Vives MC, Rubio L, Lopez C, Navas-Castillo J, Albiach-Marti MR, Dawson WO, Guerri J, Flores R, Moreno P (1999) The complete genome sequence of the major component of a mild citrus tristeza virus isolate. J Gen Virol 80(3):811–816

Vives MC, Rubio L, Sambade A, Mirkov TE, Moreno P, Guerri J (2005) Evidence of multiple recombination events between two RNA sequence variants within a Citrus tristeza virus isolate. Virology 331(2):232–237

White KA, Morris TJ (1994a) Nonhomologous RNA recombination in tombusviruses: generation and evolution of defective interfering RNAs by stepwise deletions. J Virol 68(1):14–24

White KA, Morris TJ (1994b). Recombination between defective tombusvirus RNAs generates functional hybrid genomes. Proc Natl Acad Sci USA 91(9):3642–3646

White KA, Morris TJ (1995) RNA determinants of junction site selection in RNA virus recombinants and defective interfering RNAs. RNA 1(10):1029–1040

White KA, Morris TJ (1999) Defective and defective interfering RNAs of monopartite plus-strand RNA plant viruses. Curr Top Microbiol Immunol 239:1–17

White KA, Nagy PD (2004) Advances in the molecular biology of tombusviruses: gene expression, genome replication, and recombination. Prog Nucleic Acid Res Mol Biol 78:187–226
Wierzchoslawski R, Bujarski JJ (2006) Efficient in vitro system of homologous recombination in brome mosaic bromovirus. J Virol 80(12):6182–6187
Wierzchoslawski R, Dzianott A, Kunimalayan S, Bujarski JJ (2003) A transcriptionally active subgenomic promoter supports homologous crossovers in a plus-strand RNA virus. J Virol 77(12):6769–6776
Worobey M, Holmes EC (1999) Evolutionary aspects of recombination in RNA viruses. J Gen Virol 80(10:2535–2543
Wu W, Blumberg BM, Fay PJ, Bambara RA (1995) Strand transfer mediated by human immunodeficiency virus reverse transcriptase in vitro is promoted by pausing and results in misincorporation. J Biol Chem 270(1):325–332
Yeh TY, Lin BY, Chang YC, Hsu YH, Lin NS (1999) A defective RNA associated with bamboo mosaic virus and the possible common mechanisms for RNA recombination in potexviruses. Virus Genes 18(2):121–128
Zhong Y, Guo A, Li C, Zhuang B, Lai M, Wei C, Luo J, Li Y (2005) Identification of a naturally occurring recombinant isolate of Sugarcane mosaic virus causing maize dwarf mosaic disease. Virus Genes 30(1):75–83
Zimmern (1988) Evolution of RNA viruses. In: Domingo JJHE, Ahlquist P (eds) RNA genetics, vol. II. CRC, Boca Raton, pp 211–240

Chapter 9
Symbiosis, Mutualism and Symbiogenesis

Marilyn J. Roossinck

Abstract Symbiosis is defined as two or more dissimilar entities living in or on one another in an intimate relationship. This definition encompasses both virus–virus and virus–host relationships. Symbiosis can be manifest as different lifestyles, from antagonistic (i.e., pathogenic) to mutualistic. Virus–virus and virus–host symbioses also manifest these different lifestyles, although the antagonistic lifestyles for virus–host relationships are the ones most studied, and hence most familiar. Studying viruses from the viewpoint of symbiosis emphasizes the relationships rather than the individuals in the partnerships. Symbiotic relationships can lead to the fusion of the entities, resulting in the formation of a new species, a process known as symbiogenesis. Plant viruses clearly have undergone repeated symbiogenesis in the evolution of the extant species, as evidenced by phylogenetic analyses, as well as a number of examples of viruses in the process of speciation.

9.1 Introduction

Symbiosis, as the term was originally coined by Frank and de Bary, requires two or more dissimilar entities living in or on one another in intimate contact (de Bary 1879). Symbiosis does not require mutualism, where both entities benefit from the relationship, which is only one of the potential lifestyles of a symbiotic relationship. Symbionts also can be commensal, where partners neither benefit nor are

Marilyn J. Roossinck
The S.R. Noble Foundation, P.O. Box 2180, Ardmore, OK 73402, USA
mroossinck@noble.org

M.J. Roossinck (ed.), *Plant Virus Evolution.*

harmed, or antagonistic, where one or more partners benefit at the expense of the other(s). Historically viruses have not been described in terms of symbiosis, but recently this concept has gained acceptance (Margulis et al. 2007; Ryan 2007; Villarreal 2007). For the most part, viruses have been considered only as antagonists of their hosts, and are generally described as pathogens.

For plant viruses, almost all of the literature to date describes pathogenic viruses of crop plants. Studies on the occurrence and biological effects of plant viruses in wild hosts are beginning to surface only now (Wren et al. 2006). Here I will discuss two aspects of plant viral symbiosis: symbiotic relationships with host plants; and virus–virus symbiosis in mixed infections. In both types of relationships symbiosis can lead to symbiogenesis, the generation of a novel species by fusion of two species (Ryan 2006).

9.2 Virus–Host Symbiosis

Viruses are obligate intracellular parasites, requiring host machinery for most of their biological processes. Plant viruses (excluding the algal viruses) are relatively small in genomic content, probably because a small virion is necessary for their systemic spread, and hence require more from their hosts than some of the large animal or bacterial viruses, which can have genomes larger than some bacteria (Raoult et al. 2004). The relationship between plant viruses and their hosts is, for the most part, thought to be antagonistic, but that is likely because they have been studied only in terms of disease. There are a growing number of viruses found in plants that could be classified as persistent and commensal. Cryptic viruses have been known for some time (Boccardo et al. 1987), and more recently endornaviruses (endogenous RNA viruses), a group of large double-stranded RNA viruses have been described from numerous plants (Fukuhara et al. 2006). These viruses apparently are not transmitted from plant to plant, but are passed from generation to generation through seed. Interestingly, they have relatives in the fungal viruses (Boccardo and Candresse 2005; Hacker et al. 2005), and it seems possible that they originated as fungal viruses that became trapped in plants during a plant–fungus endophytic interaction (Roossinck 1997). No phenotype has been ascribed to the presence of these viruses, but since it is difficult to cure plants of them, they may express a subtle phenotype that is not apparent. Totiviruses, another family of fungal viruses, may also infect plants (Martin et al. 2006). Mutualistic viruses have not been described in plants, but they clearly are found in animal and insect virus systems (Rossignol et al. 1985; Webb 1998; Turnbull and Webb 2002; Moran et al. 2005; Stasiak et al. 2005). Viruses are also involved in more complex symbiotic relationships that involve plants. The mutualistic endophytic fungus *Curvularia protuberata* requires a virus to confer thermal tolerance to plants (Márquez et al. 2007). Plant viruses can also provide a benefit for their insect vectors (Belliure et al. 2005). When the field of plant virology overcomes the bias for pathogenic viruses of crop plants it is likely that many more examples will be found.

It is useful to think of viruses in terms of symbionts rather than to think of plants and viruses as independent entities, because this focuses on the interactions. Although experimental studies usually strive to bring everything to its simplest form, in nature all life exists in highly interactive and interdependent communities (see Chap. 2).

9.3 Virus–Host Symbiogenesis

Symbiogenesis was first recognized as an alternative model for evolution in the early twentieth century (Carrapiço and Rodrigues 2005). The evolution of the eukaryotic cell clearly involved symbiogenic events in the incorporation of organelles, but virus–host symbiogenesis may be much more important than previously recognized. Footprints of viral genes are found throughout most genomes that have been sequenced, including those of plants (Mette et al. 2002; Staginnus and Richert-Pöggeler 2006; also see Chap. 4), and viruses have probably played important roles in the evolution of their hosts, including the development of flowering plants (reviewed in Villarreal 2005). While reports of viruses in vascular plants are abundant, especially viruses inducing acute infections, reports of viruses in primitive plants such as mosses and liverworts are lacking. One example exists of viruses found in hornworts, and these are related to viruses from crop plants (Okuno et al. 2003). The deficiency may be due to the lack of looking rather than a true absence of viruses in these plants, but if acute disease-causing infections were common in primitive plants it is likely that they would have been noticed. The evolution of vascular plants coincides with a huge expansion in genome size, and this could be due to massive colonization by genetic parasites, including viruses. The evolution of angiosperms (flowering plants) represents an explosion of new species, and coincidently, the number of viruses that can infect flowering plants is vastly larger than for other plants (Villarreal 2005). It is an intriguing possibility that the two are linked.

Virus–host symbiogenesis can also result in virus speciation through the acquisition of new genes from the host by the virus. This has been documented in at least one case (Mayo and Jolly 1991), and has been seen in experimental evolution studies involving virus infections of transgenic plants (Allison et al. 1990, 1996). The movement proteins of plant viruses may have been acquired from their hosts, as these proteins show an extraordinary level of diversity in viruses whose replicase genes are closely related (Melcher 2000), and plants express proteins that are functionally very similar to plant virus movement proteins (Lucas et al. 1995; Xoconostle-Cázares et al. 1999).

9.4 Virus–Virus Symbiosis

Mixed infections of plant viruses are common and well documented. To be considered symbiotic, however, there must be evidence of interaction. Plant virus synergy, first described in the 1950s (Rochow and Ross 1955), is a well-known example of plant virus symbiosis. In synergy, the disease symptoms are enhanced by the mixed infection. The most well documented examples of synergy involve potyviruses, and result in enhanced replication of the partner virus, while the potyvirus replication is not affected (Vance 1991; Pruss et al. 1997). At least in some cases this was shown to involve suppression of post-transcriptional gene silencing by the potyvirus (Anandalakshmi et al. 1998). Synergy occurs among DNA viruses as well as RNA viruses of plants, and has been documented for the geminiviruses (Fondong et al. 2000; Pita et al. 2001), and between an RNA virus and a DNA virus (Hii et al. 2002). Synergy can result in enhanced replication of one, both or neither partner (reviewed in Zhang et al. 2001).

Support of satellite RNAs or viruses is another form of symbiosis that is found in plant viruses. The helper virus is responsible for replication and dissemination of the satellite element. Strictly defined, satellites do not provide any essential functions for the helper virus, but they can alter the symptom phenotype by either attenuating or exacerbating symptoms (Xu and Roossinck 2001; Simon et al. 2004). In general, these elements have an antagonistic affect on the replication of the helper virus.

A number of plant viruses exhibit another form of symbiosis called interdependence. These symbiotic interactions are often obligate. They can be required for vector transmission, as in the umbravirus/luteovirus symbioses (Gibbs 1995) and the rice tungro disease viruses (Hull 2002), or in establishment of systemic infection, as with *Pepper veinal mottle virus* and *Potato virus Y* (Marchoux et al. 1993) and other potyvirus combinations, or with *Pepper mottle virus* and *Cucumber mosaic virus* (CMV; Guerini and Murphy 1999).

9.5 Virus–Virus Symbiogenesis

The role of symbiogenesis in the evolution of RNA viruses of plants has been reviewed recently (Roossinck 2005). When symbiotic viruses exchange genomic elements a new viral species is formed. The results of these symbiogenic events are usually seen in phylogenetic analyses, where different portions of the viral genomes yield trees that are not congruent (Morozov et al. 1989; Roossinck 2002; Morozov and Solovyev 2003). This has been termed "modular evolution" (Botstein 1980). The modular nature of plant virus proteins means that viruses in symbiotic relationships in mixed infections have been frequently reassorted and recombined to form new species. This gives plant viruses an enormous level of flexibility and a capacity for very rapid evolutionary changes. A number of recently described examples of new RNA virus species likely formed through symbiogenesis include *Poinsettia latent virus*, which appears be a recombinant between a polerovirus and a sobemovirus (aus dem Siepen et al. 2005); *Bean leafroll virus* and *Sugarcane yellow leaf virus*, the products of recombination events between luteo-like and polero-like viruses (Moonan et al. 2000; Domier et al. 2002); and *Bean distortion mosaic virus*, the result of both reassortment and recombination between *Peanut stunt virus* and

CMV (White et al. 1995; Roossinck, unpublished data); and the tobravirus strains I6 and N5, which resulted from the reassortment of elements from *Tobacco rattle virus* and *Pea early browning virus* (Robinson et al. 1987). Experimental evolution has also yielded new symbiogenic viruses that are the result of recombination between *Tomato aspermy virus* and CMV (Aaziz and Tepfer 1999; de Wispelaere et al. 2005).

Symbiogenesis also has been involved in the evolution of plant DNA viruses. A great deal of the emergence of new geminiviruses has been attributed to reassortment and recombination (Rojas et al. 2005; Seal et al. 2006; also see Chap. 3). Reassortment has probably been important in the evolution of the extant *Banana bunchy top virus* group as well (Hu et al. 2007). Even viroids have not escaped the role of symbiogenesis in their evolution, and recent analyses of numerous *Peach latent mosaic viroid* isolates show clear evidence of recombination (Hassen et al. 2007).

9.6 Conclusions

Considering virus–virus and virus–host interactions in terms of symbiosis and symbiogenesis provides a framework to emphasize the interactions that shape evolution. For the most part, only the antagonistic relationships in virus–host symbioses have been explored, while more mutualistic relationships have been described for virus–virus interactions. Symbiotic and symbiogenic interactions, which affect the evolution of all life on earth, are only beginning to be studied. Theoretical frameworks are still being developed. Viruses, which can adapt rapidly to new environments, provide an ideal experimental system for understanding the complex interactions in communities of life, the way life exists in nature.

Acknowledgement The author is supported by the Samuel Roberts Noble Foundation.

References

Aaziz R, Tepfer M (1999) Recombination between genomic RNAs of two cucumoviruses under conditions of minimal selection pressure. Virology 263:282–289

Allison R, Thompson G, Ahlquist P (1990) Regeneration of a functional RNA virus genome by recombination between deletion mutants and requirement for cowpea chlorotic mottle virus 3a and coat genes for systemic infection. Proc Natl Acad Sci USA 87:1820–1824

Allison RF, Schneider WL, Greene AE (1996) Recombination in plants expressing viral transgenes. Semin Virol 7:417–422

Anandalakshmi R, Pruss GJ, Ge X, Marathe R, Mallory AC, Smith TH, Vance VB (1998) A viral suppressor of gene silencing in plants. Proc Natl Acad Sci USA 95:13079–13084

aus dem Siepen M, Pohl JO, Koo B-J, Wege C, Jeske H (2005) Poinsettia latent virus is not a cryptic virus, but a natural polerovirus-sobemovirus hybrid. Virology 336 240–250

Belliure B, Janssen A, Maris PC, Peters D, Sabelis MW (2005) Herbivore arthropods benefit from vectoring plant viruses. Ecol Lett 8:70–79

Boccardo G, Candresse T (2005) Complete sequence of the RNA1 of an isolate of *White clover cryptic virus 1*, type species of the genus *Alphacryptovirus*. Arch Virol 150:399–402

Boccardo G, Lisa V, Luisoni E, Milne RG (1987) Cryptic plant viruses. Adv Virus Res 32:171–214

Botstein D (1980) A theory of modular evolution for bacteriophages. Ann N Y Acad Sci 354:484–490

Carrapiço F, Rodrigues T (2005) Symbiogenesis and the early evolution of life. Proc SPIE 5906: R1–R4

de Bary HA (1879) Die Erscheinung der Symbiose. de Bary, Strasburg

de Wispelaere M, Gaubert S, Trouilloud S, Belin C, Tepfer M (2005) A map of the diversity of RNA3 recombinants appearing in plants infected with *Cucumber mosaic virus* and *Tomato aspermy virus*. Virology 331:117–127

Domier LL, McCoppin NK, Larsen RC, D'Arcy CJ (2002) Nucleotide sequence shows that *Bean leafroll virus* has a *Luteovirus*-like genome organization. J Gen Virol 83:1791–1798

Fondong VN, Pita JS, Rey MEC, de Kochko A, Beachy RN, Fauquet CM (2000) Evidence of synergism between African cassava mosaic virus and a new double-recombinant geminivirus infecting cassava in Cameroon. J Gen Virol 81:287–297

Fukuhara T, Koga R, Aioki N, Yuki C, Yamamoto N, Oyama N, Udagawa T, Horiuchi H, Miyazaki S, Higashi Y, Takeshita M, Ikeda K, Arakawa M, Matsumoto N Moriyama H (2006) The wide distribution of endornaviruses, large double-stranded RNA replicons with plasmid-like properties. Arch Virol 151:995–1002

Gibbs M (1995) The luteovirus supergroup: rampant recombination and persistent partnerships. In: Gibbs AJ, Calisher CH, García-Arenal F (eds) Molecular basis of virus evolution. Cambridge University Press, Cambridge, pp 351–368

Guerini MN, Murphy JF (1999) Resistance of *Capsicum annuum* 'Avelar' to pepper mottle potyvirus and alleviation of this resistance by co-infection with cucumber mosaic cucumovirus are associated with virus movement. J Gen Virol 80:2785–2792

Hacker CV, Brasier CM, Buck KW (2005) A double-stranded RNA from a *Phytophthora* species is related to the plant endornaviruses and contains a putative UDP glycosyltransferase gene. J Gen Virol 86:1561–1570

Hassen IF, Massart S, Motard J, Roussel S, Parisi O, Kummert J, Fakhfakh H, Marrakchi M, Perreault JP, Jijakli MH (2007) Molecular features of new Peach latent mosaic viroid variants suggest that recombination may have contributed to the evolution of this infectious RNA. Virology 360:50–57

Hii G, Pennington R, Hartson S, Taylor CD, Lartey R, Williams A, Lewis D Melcher U (2002) Isolate-specific synergy in disease symptoms between cauliflower mosaic and turnip vein-clearing viruses. Arch Virol 147:1371–1384

Hu J-M, Fu H-C, Lin C-H, Su H-J, Yeh H-H (2007) Reassortment and concerted evolution in *Banana bunchy top virus* genomes. J Virol 81(4):1746–1761

Hull R (2002) Matthews' plant virology. Academic, San Diego

Lucas WJ, Bouché-Pillon S, Jackson DP, Nguyen L, Baker L, Ding B, Hake S (1995) Selective trafficking of KNOTTED1 homeodomain protein and its mRNA through plasmodesmata. Science 270:1980–1983

Marchoux G, Delecolle B, Selassie, KG (1993) Systemic infection to tobacco by pepper veinal mottle potyvirus (PVMV) depends on the presence of potato virus y (PVY). J Phytopathol 137:283–292

Margulis L, Hall J, McFall-Ngai M (2007) Our discipline comes of age Symbiosis 44(1):ii–iv

Márquez, LM, Redman, RS, Rodriguez, RJ, Roossinck MJ (2007) A virus in a fungus in a plant – three way symbiosis required for thermal tolerance. Science 315:513–515

Martin RR, Tzanetakis IE, Sweeney M, Wegener L (2006) A virus associated with blueberry fruit drop disease. Acta Hort 715:497–501

Mayo MA, Jolly CA (1991) The 5′-terminal sequence of potato leafroll virus RNA: evidence of recombination between virus and host RNA. J Gen Virol 72:2591–2595

Melcher U (2000) The '30K' superfamily of viral movement proteins. J Gen Virol 81:257–266

Mette MF, Kanno T, Aufsatz W, Jakowitsch J, van der Winden J, Matzke MA, Matzke AJM (2002) Endogenous viral sequences and their potential contribution to heritable virus resistance in plants. EMBO J 21(3):461–469

Moonan F, Molina J, Mirkov TE (2000) *Sugarcane yellow leaf virus*: an emerging virus that has evolved by recombination between luteoviral and poleroviral ancestors. Virology 269:156–171

Moran NA, Degnan PH, Santos SR, Dunbar HE, Ochman H (2005) The players in a mutualistic symbiosis: insects, bacteria, viruses, and virulence genes. Proc Natl Acad Sci USA 102(47):16919–16926

Morozov SY, Dolja VV, Atabekov JG (1989) Probable reassortment of genomic elements among elongated RNA-containing plant viruses. J Mol Evol 29:52–62

Morozov SY, Solovyev AG (2003) Triple gene block: modular design of a multifunctional machine for plant virus movement. J Gen Virol 84:1351–1366

Okuno K, Hama T, Takeshita M, Furuya N, Takanami Y (2003) New potyvirus isolated from *Cryptotaenia japonica*. J Gen Plant Pathol 69:138–152

Pita JS, Fondong VN, Sangaré A, Otim-Nape GW, Ogwal S, Fauquet CM (2001) Recombination, pseudorecombination and synergism of geminiviruses are determinant keys to the epidemic of severe cassava mosaic disease in Uganda. J Gen Virol 82:655–665

Pruss G, Ge X, Shi XM, Carrington JC, Vance VB (1997) Plant viral synergism: the potyviral genome encodes a broad-range pathogenicity enhancer that transactivates replication of heterologous viruses. Plant Cell 9:859–868

Raoult D, Audic S, Robert C, Abergel C, Renesto P, Ogata H, LaScola B, Suzan M, Claverie J-M (2004) The 1.2 megabase genome sequence of mimivirus. Science 306:1344–1350

Robinson DJ, Hamilton WDO, Harrison BD, Baulcombe DC (1987) Two anomalous tobravirus isolates: evidence for RNA recombination in nature. J Gen Virol 68:2551–2561

Rochow WF, Ross AF (1955) Virus multiplication in plants doubly infected with potato virus X and Y. Virology 1:10–27

Rojas MR, Hagen C, Lucas WJ, Gilbertson RL (2005) Exploiting chinks in the plant's armor: Evolution and emergences of geminiviruses. Annu Rev Phytopathol 43:361-394

Roossinck MJ (1997) Mechanisms of plant virus evolution. Annu Rev Phytopathol 35:191–209

Roossinck, MJ (2002) Evolutionary history of *Cucumber mosaic virus* deduced by phylogenetic analyses. J Virol 76(7):3382–3387

Roossinck MJ (2005) Symbiosis *versus* competition in the evolution of plant RNA viruses. Nat Rev Microbiol 3(12):917–924

Rossignol PA, Ribeiro JMC, Jungery M, Turell MJ, Spielman A, Bailey CL (1985) Enhanced mosquito blood-finding success on parasitemic hosts: evidence for vector-parasite mutualism. Proc Natl Acad Sci USA 82:7725–7727

Ryan FP (2006) Genomic creativity and natural selection: a modern synthesis. Biol J Linn Soc 88:655–672

Ryan FP (2007) Viruses as symbionts. Symbiosis 44:11–21

Seal SE, vandenBosch F, Jeger, MJ (2006) Factors influencing begomovirus evolution and their increasing global significance: implications for sustainable control. Crit Rev Plant Sci 25:23–46

Simon AE, Roossinck MJ, Havelda Z (2004) Plant virus satellite and defective interfering RNAs: new paradigms for a new century. Annual Rev Phytopathol 42:415–437

Staginnus C, Richert-Pöggeler KR (2006) Endogenous pararetroviruses: two-faced travelers in the plant genome. Trends Plant Sci 11(10):485–491

Stasiak K, Renault S, Federici BA, Bigot Y (2005) Characteristics of pathogenic and mutualistic relationships of ascoviruses in field populations of parasitoid wasps. J Insect Physiol 51:103–115

Turnbull M, Webb B (2002) Perspectives on polydnavirus origins and evolution. Adv Virus Res 58:203–254

Vance VB (1991) Replication of Potato virus X RNA is altered in coinfections with Potato virus Y. Virology 182:486–494

Villarreal LP (2005) Viruses and the evolution of life. American Society for Microbiology, Washington
Villarreal LP (2007) Virus-host symbiosis mediated by persistence. Symbiosis 44:1–9
Webb BA (1998). Polydnavirus biology, genome structure, and evolution. In: Miller LK, Ball LA (eds) The insect viruses. Plenum, New York, pp 105–139
White PS, Morales FJ, Roossinck MJ (1995) Interspecific reassortment in the evolution of a cucumovirus. Virology 207:334–337
Wren JD, Roossinck MJ, Nelson RS, Sheets K, Palmer MW, Melcher U (2006) Plant virus biodiversity and ecology. PLoS Biol 4(3):e80
Xoconostle-Cázares B, Xiang Y, Ruiz-Medrano R, Wang H-L, Monzer J, Yoo B-C, McFarland KC, Franceschi VR, Lucas WJ (1999) Plant paralog to viral movement protein that potentiates transport of mRNA into the phloem. Science 283:94–98
Xu P, Roossinck MJ (2001) Satellites of plant viruses. In: Encyclopedia of life sciences. Macmillan, London
Zhang X-S, Holt J, Colvin J (2001) Synergism between plant viruses: a mathematical analysis of the epidemiological implications. Plant Pathol 50:732–746

Chapter 10
Methods for Analyzing Viral Evolution

Marcos Pérez-Losada, Megan Porter, and Keith A. Crandall(✉)

Abstract Phylogenetic approaches are central to the study of plant virus evolution and coevolution with plant hosts. Phylogenetics is a field that is rapidly advancing and spans the population/species boundary to include examining relationships among species as well as population dynamics and genetic associations within species. With a strong phylogenetic framework, a variety of exciting questions can be addressed relative to the evolutionary history of plant viruses. In this chapter, we outline the basic approaches for estimating phylogenetic or evolutionary histories for a set of genetic sequences and then explore a variety of approaches to test hypotheses concerning viral population dynamics, selection pressures, recombination, genetic diversity, and population growth.

Keith A. Crandall,
Department of Biology, Brigham Young University, Provo
UT 84602-5255, USA
keith_crandall@byu.edu

M.J. Roossinck (ed.), *Plant Virus Evolution.*

10.1 Introduction

Phylogenetic approaches are essential for studying the diversity, origin, and distribution of plant viruses. Phylogenies, or evolutionary histories, provide insights into key innovations that afford pathogens the ability to spread and attack particular hosts. They are essential for determining what a particular pathogen might be, where it came from, and how it evolves to infect a particular host or escape attempts at eradication. Phylogenetic approaches are also being developed to examine population dynamics, to partition historical effects from current effects on population structure, and to estimate gene flow, directionality of migration, and phylogeographic relationships of unique genotypes. Yet these phylogenetic approaches are often complicated and are continually being revised and further developed. In this chapter, we review some of the basic approaches, including some of the population-genetic approaches to study plant viral evolution (Fig. 10.1). As we do so, we point the reader to some software that we have found useful in our analyses of viral data. There is, of course, a plethora of software available. For a comprehensive summary of most phylogenetic software utilities, we refer readers to the Web site of Joseph Felsenstein, who has diligently compiled such a summary with links to a wide variety of phylogenetic software packages: http://evolution.genetics.washington.edu/phylip/software.html.

10.2 Alignment Strategies

Any phylogenetic or population study of sequence data usually begins with a multiple sequence alignment (MSA) of homologous molecules is very clear, what about (those molecules with common characters due to shared ancestry) not simply common in character state. Since "alignment strategies" are the first point of our review, we will simply assume that the sequences of interest descended from a common ancestor; however, sequence homology must always be assessed before and after estimating an alignment. A MSA is a hypothesis of homology for each nucleotide or amino acid AA position in the data. For closely related taxa (e.g., clones from the same strain), highly conserved gene regions (e.g., stems in ribosomal genes), or protein-coding genes (e.g., housekeeping genes), the estimation of a MSA can be trivial and established by visual inspection. But at deeper phylogenetic levels or when working with rapidly evolving genes (e.g., viral sequences), alignment can be far from trivial and insertion and deletion events (indels or gaps) must be postulated. Inferring a MSA of AAs is computationally easier than the alignment of nucleotides because the AA alphabet is composed of 20 characters, while DNA only has four. Thus, the "signal-to-noise ratio" is much better in AA rather than nucleotide sequences (not to mention there are two thirds characters fewer to align for AAs for a given sequence). Therefore, when protein-coding genes are analyzed, the problem of inferring positional homology can be simplified by first translating the DNA sequences into AAs, aligning the resulting

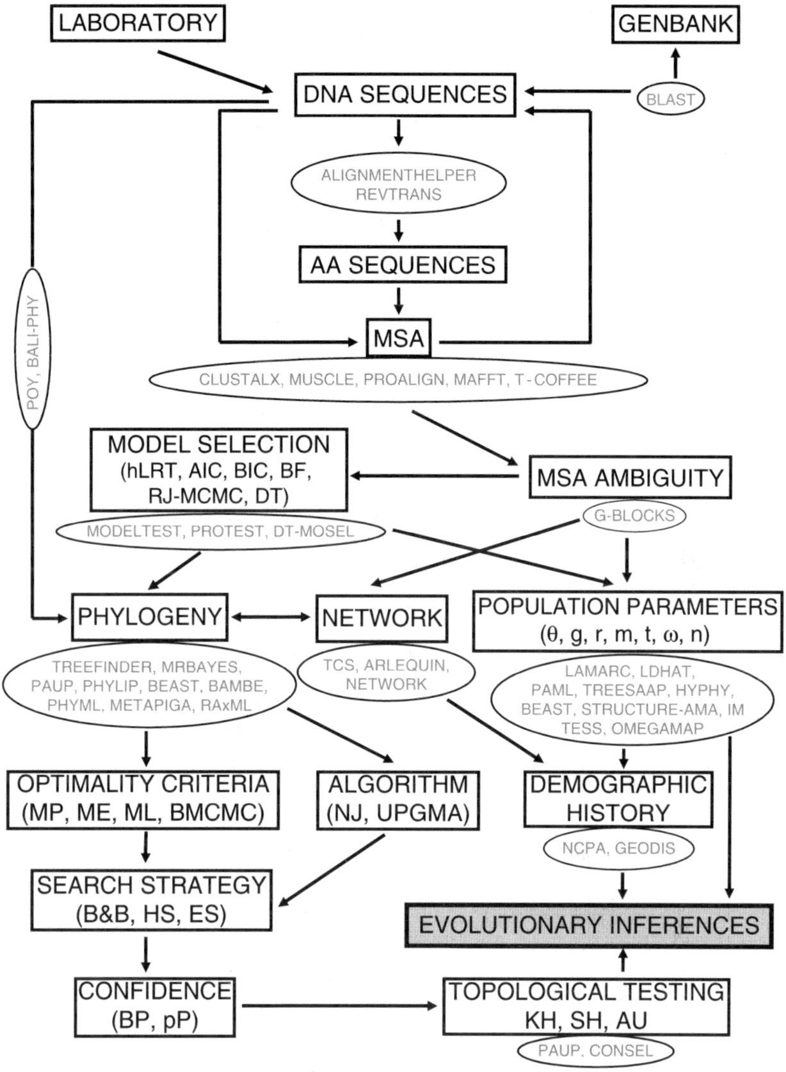

Fig. 10.1 Flow chart showing the application of various analytical approaches applied to molecular data for inferring viral population dynamics. *B&B* branch and bound, *HS* heuristic search, *ES* exhaustive search, *BP* bootstrap proportion, p^P posterior probability. Other abbreviations are explained in the main text

peptide sequences, and then converting them back to nucleotides. Computer programs such as AlignmentHelper (http://inbio.byu.edu/faculty/dam83/cdm) and RevTrans (Wernersson and Pedersen 2003) can perform this task. Even if the alignment is straightforward, coding sequences must always be aligned using a sequence editor that is capable of toggling between AA and nucleotides to be sure that the appropriate reading frame is maintained; otherwise, errors can jeopardize subsequent analyses (e.g., tests of adaptive selection). Popular sequence editors are MacClade (Maddison and Maddison 2000), Se-Al (http://evolve.zoo.ox.ac.uk/software.html), BioEdit (http://www.mbio.ncsu.edu/BioEdit/bioedit.html), and Squint (http://www.bioinformatics.org.nz/).

10.2.1 What Multiple Sequence Alignment Method To Choose?

This is a difficult question, given the variety of methods for assembling a MSA. In fact, all of the available methods use approximations (heuristics). Moreover, observed performance differences in comparative analyses (see later) usually emerge as average estimates; hence, approaches that work well for a certain gene or protein family may not work as well for a different one. Therefore, as a standard procedure, one should use multiple alignment approaches and parameter sets and carefully inspect the results (reviewed in Duret and Abdeddaim 2000; Notredame 2002). Here we will review different global alignment procedures (i.e., for sequences related over their whole length) to perform MSA. Alignment is one of the most important but ironically underappreciated and neglected aspects of sequence analysis (Crandall et al. 2005; Felsenstein 2004); hence, we will endeavor to explain here the strategies underlying some of the most commonly used algorithms, as well as their strengths and caveats.

10.2.1.1 Progressive Algorithms

Progressive alignment algorithms are by far the most widely used because of their speed, simplicity, and efficiency. The basic strategy of these methods is first to estimate a tree and then to construct a pairwise alignment of the subtrees found at each internal node. More sophisticated algorithms (e.g., iterative algorithms) also use this basic strategy in the initial or final steps of their routines. The most frequently used progressive algorithm is the one implemented in ClustalW (Thompson et al. 1994) and its window interface ClustalX (Thompson et al. 1997). The basic MSA algorithm consists of three main stages: (1) all pairs of sequences are aligned separately and an uncorrected distance matrix is calculated (Sainudiin et al. 2005); (2) a guide tree (neighbor-joining, NJ, tree; Saitou and Nei 1987) is calculated from the distance matrix; (3) the sequences are progressively aligned (profile alignment) according to the branching order in the guide tree. The main caveat of this strategy is that any misaligned regions created early in the process cannot be corrected later as new sequences are added. Benchmarking tests (i.e., databases of reference structural alignments used to assess performance of MSA methods) carried out in BALiBASE (Thompson et al. 1999b) showed that ClustalW performs better when the phylogenetic tree is relatively dense without any obvious outliers (Thompson et al. 1999a). Long insertions or deletions can also be problematic owing to the intrinsic limitation of the implemented affine penalty scheme. ClustalX includes quality analysis tools, which allow for the identification of problematic regions and realigning by adjusting the gap penalties (i.e., refinement). The application TuneClustalX (http://homepage.mac.com/barryghall/Software.html) run in conjunction with ClustalX can aid in constructing a better alignment. Clustal can align both nucleotides and AAs. For the latter, BLOSUM, PAM, GONNET, and identity matrixes can be implemented. It can also use Swiss-Prot secondary structure information.

10.2.1.2 Consistency-Based Algorithms

An improved progressive strategy is implemented in T-Coffee (Notredame et al. 2000) where sequences are aligned in a progressive manner but using a consistency-based objective function that minimizes potential errors in the early stages of the alignment assembly. It works as follows: first it generates two primary libraries of pairwise global ClustalW and local LALIGN (FASTA package; Pearson and Lipman 1988) alignments and assigns weights to each pair; then both libraries are combined in a new primary library by a process of addition and weights are reestimated; a position-specific substitution matrix (extended library) is then created by examining the consistency of each pair of residues with residue pairs from all of the other alignments; this new library (and weights) is finally resolved by using a progressive alignment strategy similar to that implemented in Clustal to give a MSA. Comparison with ClustalW using the BALiBASE database indicates that T-Coffee is significantly more accurate, but about 2 times slower. A novel method (3DCoffee) has been published (O'Sullivan et al. 2004) that combines protein sequences and 3D-structures in order to generate high-quality MSA.

10.2.1.3 Iterative Algorithms

The strategy here is to produce an alignment using the progressive approach and then refine it through a series of cycles (iterations) until no more improvements can be made. Examples of methods implementing this strategy are MUSCLE (Edgar 2004) and MAFFT (Katoh et al. 2002), the fastest known algorithms. MUSCLE generates a refined alignment in three basic steps. In an initial stage (draft progressive), it produces a MSA1 quickly (speed is emphasized over accuracy) using uncorrected distances and the unweighted pair group method using arithmetic mean (UPGMA; TREE1) method. In a second stage (improved progressive), an improved MSA2 is generated by reestimating a new guide tree (TREE2) using the Kimura 2-parameter (K2P) distance, which corrects for multiple substitutions per site. In a final stage (refinement), TREE2 is divided into two subtrees, for which two profiles are computed. A new MSA is then produced by realigning the two profiles (MSA3). If MSA3 has a better score than MSA2 (as indicated by the log-expectation function implemented), then the new alignment is kept; otherwise, it is discarded. The refinement ends when convergence is reached. MUSCLE implements three different protein profile scoring functions: log-expectation score (it gives the best results and is also the only option for nucleotides), and sum of pairs scores using either the PAM200 matrix or the VTML240 matrix. MAFFT (Katoh et al. 2005) implements a similar strategy, but offers a multiple array of algorithms for the progressive and refinement processes that implement a Fourier transform approximation and include local or global pairwise alignment information. Moreover the user can choose among different AA scoring matrixes: BLOSUM (the most accurate), PAM, and JTT.

Notredame et al. (2000), Edgar (2004), and Katoh et al. (2005) compared the performance of these four programs using multiple benchmark alignment databases. The results can be summarized as follows for speed MUSCLE > MAFFT > ClustalW > T-Coffee, and for accuracy MAFFT > MUSCLE > T-Coffee > ClustalW. However, these relative comparisons must be interpreted with caution because the results are averaged over large numbers of tests and did not include the same (or most recent) versions of the software tested.

10.2.1.4 Hidden Markov Methods Based Algorithms

Hidden Markov Methods (HMMs) describe the MSA in a statistical context, using a Bayesian approach (see phylogenetic inference). From a formal point of view they are very attractive because they assign a posterior probability value to particular MSAs and sites, which allows a statistical evaluation of alternative alignments and identification of unreliable alignment regions, but it has the burden of being very computationally intense (i.e., limited to small datasets of approximately 25 taxa). ProAlign (Löytynoja and Milinkovitch 2003) is an example of this approach that combines a pair of HMMs, a progressive algorithm, and an evolutionary model (see model selection) describing the nucleotide or AA substitution process and the occurrence of gaps. This combination allegedly improves the accuracy of MSA and our understanding of the history and function of the sequences. Comparative performance tests with ClustalW using simulated data and the BALiBASE database indicated that ProAlign was more accurate, albeit slower, for the aligning of nucleotide data.

10.2.2 Treating Highly Divergent Segments of the Alignment

Since not all gene regions evolve at the same rate (e.g., stem compared with loop regions of ribosomal RNA), some parts of the MSA are reasonably conserved, whereas others are very divergent and full of gaps; hence, positional homology cannot always be precisely determined. In such cases, some authors (e.g., Gatesy et al. 1993; Swofford et al. 1996) recommend deleting those regions from subsequent analyses because they can be misleading. This is usually done in an arbitrary way, which makes the final alignment irreproducible. Some of the methods described earlier (e.g., ClustalX) can help to identify those poorly aligned regions, but more objective approaches such as Gblocks (Castresana 2000) have been described for removing very divergent regions or gap positions from an alignment of DNA or protein sequences. Gblocks selects ambiguous blocks from the MSA according to a simple set of alignment position features, including minimum number of sequences for a conserved and a flank position, maximum number of contiguous nonconserved positions, minimum length of a block, and allowed gap positions.

10.3 Model Selection: Beyond Kimura 2-Parameter

Model specification is a critical issue in molecular phylogenetics and population inference, as the model implemented (or lack thereof) affects most downstream analyses, including estimates of phylogeny, substitution rates, bootstrap values, posterior probabilities, tests of the molecular clock (Buckley 2002; Buckley and Cunningham 2002; Pupko et al. 2002; Suzuki et al. 2002), and estimates of key population parameters such as genetic diversity, recombination, growth, and natural selection (Yang et al. 2000; McVean et al. 2002; Posada et al. 2002; Kuhner et al. 2005). A model of evolution simply defines in statistical terms the probability of changing character states. So, for nucleotide sequence data, for example, a model defines the probability of changing states from an A to a T, C, or G or remaining an A and likewise for the rest of the nucleotides. The difficulty is then how to choose among the wide variety of models possible. In fact, the ability to select models within a rigorous statistical framework is one of the many advantages of explicitly model-based methods. Yet many researchers using these methods still rely on program-defined default parameter values and models, even though numerous studies have shown that phylogenetic methods are less accurate or become inconsistent when the model of evolution is misspecified (Felsenstein 1978; Huelsenbeck and Hillis 1993; Penny et al. 1994; Gaut and Lewis 1995; Sullivan and Swofford 1997; Bruno and Halpern 1999; Yang and Bielawski 2000). Regardless of data type (i.e., nucleotide or AA sequences), identifying the most appropriate model is essential to increasing the accuracy, consistency, and confidence of phylogenetic analyses and population parameter estimation. Therefore, how do we choose the best-fit model for our sequence data? This issue is usually assessed within a phylogenetic framework and has received a great deal of attention recently, leading to a suite of new methodologies (see Posada and Buckley 2004; Sullivan and Joyce 2005). In this section, we briefly review the available methods of model selection for both nucleotide and AA data, so that researchers studying viral evolution can move beyond the K2P, or default parameter, perspective.

10.3.1 Nucleotide Data

Although model choice is a crucial part of phylogeny estimation, selecting a set from the 203 "standard" time-reversible models of nucleotide substitution is not easy, especially when most model selection methods are limited to a subset of these

(Huelsenbeck et al. 2004; Posada and Buckley 2004). Model choice will be complicated further by increasing complexity, as parameters reflecting new information on nucleotide substitution processes are added to candidate models. Furthermore, model selection is moving towards using confidence sets of models for phylogeny estimation by estimating a tree for each candidate model in a 95% confidence set and then building a consensus tree using model weights (Akaike weights, Bayesian information criterion, BIC, weights, or model likelihoods from Bayesian analyses) as tree weights (Posada and Buckley 2004). Although model selection is a crucial step in phylogeny estimation, there is "no substitute for careful thinking and common sense reasoning" when selecting the model of evolution (Browne 2000).

10.3.1.1 Hierarchical Likelihood Ratio Tests

Huelsenbeck and Crandall (1997), Frati et al. (1997), and Sullivan et al. (1997), all proposed a method of model selection involving successive pairwise comparisons of nested models using hierarchical likelihood ratio tests (hLRTs) to determine the best-fit model at each step. These pairwise comparisons are made in a specific sequence until a model is found that cannot be rejected. This suggested methodology was later implemented in the program Modeltest (Posada and Crandall 1998), which works in conjunction with the commonly used phylogenetic analysis program PAUP* (Swofford 2002) to test 56 models of evolution. Since then, hLRTs have become the most widely used strategy of model selection. However, while hLRTs are a huge improvement over arbitrary model choice (or no choice at all), recent studies have shown that hLRTs have some undesirable characteristics. First, hLRT methods attempt to find the model that best fits the data under the assumption that at least one of the models compared is correct, even though all candidate models will be misspecified (i.e., the "true model" is unknown). Furthermore, hLRTs perform multiple tests with the same data, which will increase the rate of false positives, and the model chosen can be affected by whether the pairwise comparisons start with either the simplest or the most complex models. The hLRT methods are also unable to accomplish model averaging or assess model selection uncertainty. Finally, hLRTs can only provide information regarding the *relative* fit of the nested alternatives, but cannot evaluate the absolute goodness of fit of the chosen model (Minin et al. 2003).

10.3.1.2 Akaike Information Criterion

To overcome some of the issues surrounding the use of hLRTs, more recent model selection methods (Molphy – Adachi and Hasegawa 1996; Modeltest – Posada and Crandall 1998) implement several estimators (Akaike information criterion, AIC, differences, Akaike weights) based on the AIC for evaluating model fit (Posada and Buckley 2004). In a phylogenetic context, the AIC is designed to choose the model that best *approximates* reality and represents the amount of information lost when using a given model to approximate the real process of nucleotide substitution (Posada and Buckley 2004). In comparison with hLRTs, AIC statistics have the advantage of being able to simultaneously compare all candidate models (nested and nonnested), assess model selection uncertainty, and allow for model-averaged parameter estimates and relative parameter importance. Furthermore, whereas hLRTs tend to favor more complex models (Burnham and Anderson 2002), the AIC includes a penalty for overparameterization (Sullivan and Joyce 2005). In addition to the original AIC, there are several derived AIC statistics used for model selection: the second-order AIC, AIC_c, should be used when sample size (n) is small compared with the number of parameters (K); because the AIC is a relative score, AIC differences can be used to rank candidate models, with larger AIC differences being less probable; Akaike weights (w) can be used for assessing model selection uncertainty by constructing 95% confidence sets of models by summing w from largest to smallest.

10.3.1.3 Bayesian Methods

Although Bayesian methods as applied to both phylogenetic and population parameter inference are relatively new, model selection and/or the estimation of the model parameters in an a priori specified model can be an integral part of these analyses. Bayesian approaches of model selection are designed to identify the true model given the data and they have several advantages over standard hLRTs, including the ability to compare nonnested sets of models and to make inferences based on the entire set of candidate models (i.e., model averaging). Additionally, Bayesian model selection methods are not dependent on a single topology or a particular set of model parameters, making the results more valid (Nylander et al. 2004). Model selection can be incorporated into this framework in several ways, including the use of posterior probability, BIC, and Bayes factors (BF). Perhaps the most common way of assessing confidence in a hypothesis (or model of evolution) within a Bayesian framework is to choose the solution with the highest posterior probability; furthermore, model uncertainty can be accounted for by ranking models according to their posterior probability and constructing a 95% credible interval by summing these probabilities. However, calculating model posterior probability can be computationally intensive. The BIC offers a more computationally feasible approach than calculating model likelihoods (Schwarz 1978). BIC statistics also allow for simultaneous comparison of multiple models.

The third method of model selection in Bayesian analyses is the use of BF. Similar to hLRT methods, BF consist of multiple pairwise comparisons of evidence provided by the data for two competing models (Kass and Raftery 1995; Raftery 1996) and are being used for model selection in phylogenetics (Suchard et al. 2001; Aris-Brosou and Yang 2002; Huelsenbeck et al. 2004; Nylander et al. 2004). Although the interpretation of BF is up to the investigator, the general guidelines state that BF scores of more than 150 are very strong evidence for a model, 20–150 is strong, 3–20 is positive, 1–3 barely worth mentioning, and if less than 1 there is evidence for the competing model (Kass and Raftery 1995; Raftery 1996). However, because BF consist of pairwise comparisons, this statistic may have some of the same issues as hLRTs. While studies with empirical data have shown BF to be useful for selecting among complex models, it is still unclear whether this statistic represents a reasonable balance between model complexity and error in parameter estimates (Nylander et al. 2004).

Another advantage of Bayesian methodology is the ability to directly obtain a model-averaged estimate of phylogeny using an algorithm that moves through both parameter and model space (Green 1995). This type of "reversible jump Markov chain Monte Carlo" (RJ-MCMC) algorithm has recently been implemented by Huelsenbeck et al. (2004). RJ-MCMC combines model selection and phylogeny in a single step, allowing for the screening of a large number of complex candidate models while performing a phylogenetic analysis. In contrast to other model selection statistics that are limited to a small set of candidate models, this method is capable of evaluating all possible time-reversible models while accounting for uncertainty in the model during phylogeny estimation. The RJ-MCMC accomplishes this feat by implementing a MCMC algorithm that jumps between models visiting each in proportion to the posterior distribution, allowing calculation of BF for any of the models and for averaging over the possible models while performing phylogeny estimation (Huelsenbeck et al. 2004).

Even with these differences, the Bayesian and likelihood approaches seem to arrive at similar results (Nylander et al. 2004). In comparative studies of model selection, Huelsenbeck et al. (2004) found that AIC, BIC, posterior probabilities, BF, and RJ-MCMC are largely concordant, either choosing the same "best" model or choosing a model within the 95% credible set of models from RJ-MCMC analyses. Thus, given similar model choice, Bayesian methods have a computational advantage.

Within the Bayesian framework it is also possible to incorporate different models for different partitions (e.g., different genes, different codon positions, or ribosomal stems versus loops) within a dataset. This can be accomplished by determining partitions a priori, estimating a model for each partition using any of the methods discussed, and using these models (either linked or unlinked) in mixed model Bayesian analyses (Ronquist and Huelsenbeck 2003). Alternatively, the number of partitions contained within a dataset can be determined during the phylogenetic analyses using a pattern-heterogeneity mixture model (Pagel and Meade 2004).

10.3.1.4 Decision Theory

Minin et al. (2003) recently proposed a performance-based method of model selection. The decision theory (DT) approach is an extension of the BIC that improves upon previous model selection methods by incorporating relative branch-length error as a measure of phylogenetic performance. This method assumes that all candidate models are wrong and instead attempts to identify the model that incurs the least risk while attempting to minimize the number of model parameters (Minin et al. 2003). Designed to choose the simplest model that minimizes relative branch-length error, models are penalized for overfitting (i.e., more complex models are penalized if simpler models perform similarly with fewer parameters). As a result, DT generally selects simpler models that provide good or better estimates of branch lengths than the complex models selected by hLRTs for the same data (Sullivan and Joyce 2005). As an extension of the BIC, the DT approach is capable of comparing all competing models simultaneously, obviating the issues related to pairwise comparisons (hLRTs, BF). A recent study comparing DT approaches with BIC, AIC, and likelihood ratio test (LRT) approaches illustrates that model choice using LRTs and the AIC results in more complex model choices, leading to significant increases in computational time without contributing to increased accuracy in phylogenetic inference (Abdo et al. 2005). Further studies comparing DT approaches with Bayesian and likelihood approaches will help decipher the similarities/differences of this philosophically different approach to model selection. However, by incorporating a performance-based penalty, this approach attempts to identify the best-fit model that also produces the best estimates of phylogeny. This model selection method is implemented in the program DT-ModSel (http://www.webpages.uidaho.edu/~jacks/DTModSel.html).

10.3.2 Amino Acid Sequences

Modeling protein evolution is a more complex task than dealing with evolution at the nucleotide level, and accordingly fewer model-based phylogenetic and population analyses are performed on AA sequences. However, with the recent availability of programs such as PhyML (Guindon and Gascuel 2003), a program capable of using AA data in a likelihood framework for fast phylogenetic reconstruction, some of these issues are being overcome, increasing the importance of model selection for phylogenetic estimation using AA sequences as well. Owing to computational and data-complexity issues, models of protein evolution are preferentially based on empirical matrices estimated from large datasets of diverse protein families, resulting in matrices of the relative rates of replacement from one AA to another. A number of these types of matrices have been calculated (Dayhoff – Dayhoff et al. 1978; JTT – Jones et al. 1992; WAG – Whelan and Goldman 2001; mtREV – Adachi and Hasegawa 1996; MtMam – Cao et al. 1998; VT – Muller and Vingron 2000; CpREV – Adachi et al. 2000; RtREV – Dimmic et al. 2002; Blosum62 – Henikoff and Henikoff 1992), and pose the same issues as selecting the best-fit model of nucleotide evolution. To deal with the issue of model selection for AA data, the program ProtTest (Abascal et al. 2005) was developed. ProtTest computes the likelihood of each of 64 candidate models of protein evolution and estimates the fit of all the candidate models using either AIC, AIC_c, or BIC. ProtTest also calculates the importance of and provides model-averaged estimates for the relevant parameters, including I (invariable sites), Γ (gamma rate distribution), and F (observed AA frequencies from data, equivalent to equilibrium frequencies) (Posada and Buckley 2004).

Although ProtTest identifies the most appropriate AA model from among the most commonly used matrices, a secondary issue is whether or not the empirical candidate models accurately reflect evolutionary processes in a wide range of proteins. Because most of the commonly used empirical matrices were calculated from large datasets representing extreme protein family diversity, the estimated relative rates of changes may be too general to fit datasets of specific gene families. To address this concern, a second approach to justifying model choice for phylogeny estimation using AA sequences is to generate gene-specific empirical matrices via the program MatrixGen (http://matrixgen.sourceforge.net). For example, if you were interested in estimating the phylogeny of the rhodopsin superfamily of G-protein coupled receptors, you would be able to use databases such as Pfam (http://pfam.wustl.edu) to obtain large sets of aligned rhodopsin sequences from which MatrixGen can calculate a number of different empirical matrices based only on sequences related to those being investigated, rather than all proteins currently characterized, providing an empirical frequency matrix specific to the particular gene of interest.

10.4 Phylogenetic Inference: Picking Trees from the Forest

10.4.1 Bifurcation-Based Methods

Although the phylogenetic reconstruction of trees depends on the alignment and the implemented model of evolution as previously discussed, there is now a new set of choices to be made, including selecting a metric for evaluating the "quality" of each tree and a method for navigating the tree space in search of the best trees. In general, phylogenetic reconstruction methods can be divided into two types, those that proceed algorithmically and those based on optimality criteria. For further understanding of these methods, the reader is referred to the many sources discussing the merits of different theoretical approaches to phylogenetic inference (Felsenstein 1981, 2004; Huelsenbeck 1995; Swofford et al. 1996; Page and Holmes 1998).

10.4.1.1 Tree Metrics

Although all phylogenetic methods are accomplished using algorithms, only with distance-based clustering methods is the "best" tree defined by the algorithmic steps used, with no exploration of the set of possible trees (i.e., the "tree space"). Distance methods condense data to the observed pairwise differences between sequences, which can be "corrected" using a model to reflect true evolutionary distances. In general, distance methods distill all of the available information from two sequences down to a single metric, losing potentially valuable information coded in the individual characters (Huson and Steel 2004). However, the distance calculation also has some advantages, i.e., distance estimates may be more robust to alignment error than site-dependent methods (Rosenberg 2005) and fast distance-based methods can be used to produce reasonably accurate starting trees for more thorough optimality-based heuristic searches, thereby considerably decreasing computational times of existing methods (see discussion below; Guindon and Gascuel 2003). The most common distance clustering methods are NJ (Saitou and Nei 1987) and UPGMA (Sokal and Sneath 1963); however, UPGMA has the methodological disadvantage of constraining branch lengths to satisfy a "molecular clock." As most datasets do not meet this assumption (Graur and Martin 2004), UPGMA can be inefficient and extremely sensitive to branch-length inequalities, producing seriously misleading results (Huelsenbeck 1995). In contrast, the NJ method does not assume a molecular clock. Simulation studies have shown NJ to perform well (Huelsenbeck 1995), serving as a good approximation for more statistical distance methods (i.e., minimum evolution and least squares; Felsenstein 2004). Furthermore, the reasonable accuracy and fast computational speed of NJ methods allow for phylogenetic inference of very large datasets (hundreds to thousands of taxa) where other methods are computationally impossible (Tamura et al. 2004).

Preferential to algorithmically constructed trees are methods where topologies are compared on the basis of a chosen criterion, with the best tree being the one that minimizes the criterion. The most common optimality criteria for evaluating trees are distance, parsimony, likelihood, and Bayesian metrics. In addition to clustering methods, distance metrics can also be used as optimality criteria for minimum evolution (ME) inference, which has been shown to be statistically consistent when used in conjunction with ordinary least-squares fitting of a metric to a tree structure (Rzhetsky and Nei 1993; Desper and Gascuel 2002). With a parsimony-based criterion, the number of changes necessary to make the data fit a given tree are counted and the tree with the lowest score (number of character changes along the tree) is chosen as best. Maximum parsimony (MP) as an optimality criterion considers only those character differences visible in a given dataset. However, the remaining criteria (likelihood and Bayesian) are calculated based on a probabilistic model of evolution, which can account for unobservable sequence variation. Maximum likelihood (ML) inference attempts to identify the topology that explains the evolution of a set of aligned sequences under a given model of evolution with the greatest likelihood (nucleotide – Felsenstein 1981; AAs – Kishino and Hasegawa 1989). Many simulation studies have identified the likelihood criterion (Felsenstein 1981) as one of the best for phylogenetic inference, citing properties of statistical consistency, robustness, the ability to compare trees within a statistical framework, and the ability to make full use of the original character matrix (reviewed in Whelan and Goldman 2001). However, as one of the most computationally intensive optimality criteria, its use is limited to smaller numbers of taxa. Although similar to ML, Bayesian inference combines the prior probability of a phylogeny with the likelihood, producing a posterior probability distribution of trees, which can be interpreted as the probability that the tree is correct (Huelsenbeck et al. 2001). Bayesian methods have risen quickly to the forefront of phylogenetics as a likelihood-based method that is able to search reasonable portions of the tree space and assess the confidence of the estimated relationships in realistic computational timeframes. Both algorithmic and optimality-criteria-based methodologies can be implemented in a number of commonly used phylogenetic programs (NJ, ME, MP, ML: PAUP – Swofford 2002; PHYLIP – http://evolution.genetics.washington.edu/phylip/phylip.html; Bayesian inference: MrBayes – Ronquist and Huelsenbeck 2003; BAMBE – Simon and Larget 2000).

10.4.1.2 Search Strategies

The theoretically ideal situation is to evaluate all possible trees on the basis of the chosen criterion in order to identify the best (i.e., exhaustive search); however, given the unfathomable number of possible trees for even small datasets, this method quickly becomes untenable. The next best option is the branch-and-bound method, which is guaranteed to find all of the optimal solutions without doing an exhaustive search. This is accomplished by keeping track of the score of the current best solution as the tree is being constructed; as branches are added, topologies suboptimal to the current best (and all related topologies) can be discarded from further analyses, reducing the number of topologies to be evaluated (Hendy and Penny 1982). Branch-and-bound methods are also severely limited by the number of taxa that can be evaluated within reasonable time limits. Therefore a number of heuristic algorithms, which sacrifice the guarantee of finding the optimal solution(s) for reduced computational time, have been developed. The most common phylogenetic heuristic search type is based on hill climbing, where an initial tree is subject to topological rearrangement. The new tree is either kept and used as the new starting tree or rejected depending on the change in tree score. The current best tree is subjected to rearrangement until the tree score can no longer be improved. This rearrangement process is then replicated many times using different starting trees and the tree score is compared among replicates to identify the best tree or set of trees. However, as the size of datasets increases, traditional hill-climbing heuristics have become computationally intractable, even for the faster MP methods. One solution to the computational bottleneck that has been explored for MP searches is a process called the "ratchet" (Nixon 1999). The ratchet can be implemented using the following steps: (1) generate a starting tree; (2) randomly perturb the dataset via random character reweighting; (3) perform branch swapping on the current tree using the new reweighted matrix, holding a single tree or a few trees; (4) return to the original dataset and perform branch swapping on the tree from step 3; (5) return to step 2 and repeat, using trees from step 4 as the new starting point (Nixon 1999). This process has been shown to move a search around the tree space much more effectively, especially for large datasets. Another approach to MP searches is direct optimization, where topologies are evaluated without first creating MSAs (Wheeler 1996; Wheeler et al. 2003; Sec. 10.4.2).

Owing to the need for parameter optimization at each step, increasing complexity in evolutionary models, and larger datasets, ML inference is the most computationally intensive method; accordingly, more focus has been placed on improving search strategies/decreasing computational time for ML heuristics in particular. One approach is implemented in the program PHYML (Guindon and Gascuel 2003), where an initial tree built using a fast distance-based method is subjected to a simple hill-climbing heuristic in which computational time is significantly improved by adjusting both tree topology and branch lengths simultaneously. This simultaneous adjustment is a compromise between speed and accuracy, and requires only a few iterations to reach an optimum. Other ML-based programs that implement fast algorithms allowing for mixed models (i.e., different models for different data partitions) and bootstrapping procedures (see later) are TREEFINDER (Jobb 2005), RAxML (Stamatakis et al. 2005), and HyPhy (Kosakovsky Pond 2005). Another recent improvement to ML heuristic approaches is the implementation of genetic algorithms (GA; Matsuda 1996; Lewis 1998; Katoh et al. 2001; Lemmon and Milinkovitch 2002). GAs are a type of evolutionary computation method where the tree space is navigated by randomly perturbing a population of trees via branch length and topology modification, obtaining better trees by recombining the perturbed trees, selecting the best tree(s), and repeating the process until an optimum is reached (Lewis 1998). The population of trees is perturbed using a set of operators that mimic processes of biological evolution (i.e., mutation, recombination, selection, and reproduction) and the trees are then combined to produce better trees by allowing trees to "reproduce" with a probability based on a value of relative fitness (Lemmon and Milinkovitch 2002). As the relative fitness of each tree is a function of the optimality score, GAs simulate natural selection and the mean score of the population of trees improves over time. The GA continues to let populations "evolve" until either a cutoff point is reached, or the populations of trees stop improving in score. The most commonly used program implementing GAs is MetaPIGA, which uses a metapopulation setting (the metaGA) relying on the interactions of two or more populations of trees (Lemmon and Milinkovitch 2002). Another recently developed program for fast ML estimation using a GA approach is GARLI (http://www.zo.utexas.edu/faculty/antisense/Download.html), which is apparently more accurate than PHYML (i.e., finds better likelihood trees) and approaches RAxML (Stamatakis et al. 2005) speeds for datasets of fewer than 1,000 sequences. RAxML seems to remain the best ML option for datasets of greater than 1,000 sequences.

Since its implementation, Bayesian inference using Metropolis-coupled MCMC (BMCMC) methods has rapidly become the favored method for phylogenetic tree reconstruction (Simon and Larget 2000; Huelsenbeck and Ronquist 2001; Drummond and Rambaut 2003; Pagel and Meade 2004). Contrary to inference using other optimality criteria, the goal of BMCMC methods is to sample the posterior probability distribution of trees contained by the tree space. BMCMC methods generate a Markov chain starting with an arbitrary set of parameter values that are updated using a stochastic proposal mechanism in each cycle, with the proposed new state accepted on the basis of a probability determined by the product of the prior ratio, the likelihood ratio, and the proposal ratio (Nylander et al. 2004). Although theoretically a Markov chain should produce a valid sample of the posterior probability distribution (Tierney 1994), one of the major issues of BMCMC analyses is determining how long to run a chain to accomplish this goal (Nylander et al. 2004). To determine whether Markov chains have approximated the targeted posterior distribution, most analyses consist of at least three independent runs started from different random sets of parameters/tree topologies run for at least 5×10^6 cycles. These independent runs are then compared to determine the convergence and mixing behavior of each analysis using programs such as Tracer (Rambaut and Drummond 2003). Convergence on similar distributions can be assessed by plotting the likelihood score over cycle number for each chain; to assess mixing, however, examination of all parameter changes relative to cycle numbers is required (Nylander et al. 2004). Further concerns lie with the appropriate choice of prior probabilities for each parameter of interest (Zwickl and Holder 2004; Yang and Rannala 2005). BMCMC methods offer several practical advantages over more traditional hill-climbing heuristic searches, including faster computational time relative to ML, simultaneous assessment of both tree and clade support, the ability to accomplish analyses incorporating mixed models for molecular and morphological partitions (Ronquist and Huelsenbeck 2003; Pagel and Meade 2004), phylogeny estimation while accounting for model uncertainty (Huelsenbeck et al. 2004), and among the most recent advantages, simultaneous alignment and phylogeny estimation (Lunter et al. 2005; Redelings and Suchard 2005).

As our increasing ability to generate large and complex datasets outpaces our ability to accomplish analyses in reasonable timeframes, the computational efficiency of phylogenetic algorithms has become a focal area for improvement. For optimality-based methods in particular, the greatest potential for improving the computational speed of analyses lies in improved algorithmic search strategies rather than in improved hardware capabilities. However, as the development of new algorithms takes time, recent efforts have also been focused on implementing parallel processing routines for a number of common programs (Janies and Wheeler 2001; Brauer et al. 2002; Schmidt et al. 2002; Ronquist and Huelsenbeck 2003). In almost all cases, parallelization provides considerable improvement in computational speeds.

10.4.1.3 Confidence Assessment

Once a phylogeny has been estimated, the next step is to assess the confidence of the estimated relationships. The nonparametric bootstrap procedure (Felsenstein 1985) is commonly used for estimating nodal support under traditional methods of phylogenetic inference and posterior probabilities are used in Bayesian inference. An alternative cladistic approach is the Bremer support (Bremer 1988; or decay index, Donoghue et al. 1992), which is performed under the MP criterion. However, we do not support the use of this method because it does not provide statistical measures of clade uncertainty and is not comparable between trees or datasets. The bootstrap procedure resamples the original dataset to create a new dataset by choosing columns of data from the original data matrix at random with replacement until a new data matrix is created that has the same sequence length as the original. Then a tree is estimated from this resampled dataset. This procedure is repeated multiple times (typically 100 times for ML and 1,000 times or more for MP, ME, and NJ) to achieve reasonable precision. Hillis and Bull (1993) showed that bootstrap proportions provide biased (i.e., they vary from branch to branch and study to study) but highly conservative estimates of the probability of correctly inferring the corresponding clades, suggesting that bootstrap proportions of 70% or greater correspond to a probability of 0.95 or greater that the clade was real under the conditions of their study. However, the bias associated with the bootstrap can become pronounced with large-scale phylogenies and can thereby reduce the accuracy of the confidence assessment (Sanderson and Wojciechowski 2000).

Posterior probabilities are the measure of confidence for Bayesian phylogenies. They have a straightforward interpretation as the probability that a particular monophyletic group is correct, but extensive debate has focused on whether and how these proportions can be meaningfully related to phylogenetic accuracy and frequentist testing (Sanderson 1995). Bayesian posterior probabilities tend to give higher support for nodes than bootstrap values, sometimes with little correlation between the two measures at corresponding nodes (Leaché and Reeder 2002). This causes disagreement on how posterior probabilities should be interpreted relative to nonparametric bootstrap proportions (Alfaro et al., 2003; Douady et al. 2003). The fact is that the methods measure different, yet complementary, features of the data; therefore, both should be estimated.

10.4.1.4 Testing Alternative Hypothesis

Frequently a topology estimated for one gene partition is in conflict with a second topology estimated from another gene partition or from the same gene partition using a different phylogenetic approach. In such cases, it is necessary to statistically test if the alternative topology is significantly different from the optimal topology. Different paired-sites tests (Felsenstein 2004) and Bayesian tests (Huelsenbeck et al. 2002) have been described for comparing trees using either the MP and ML scores or posterior probabilities (Sinclair et al. 2005). The distinction between these tests comes in the clarification of whether one is comparing a priori (i.e., all the phylogenies being tested are independent of the results of the phylogenetic analysis) or a posteriori (i.e., at least one phylogeny in the test is derived from the phylogenetic analysis) hypotheses and the number of trees compared. Bayesian methods assess the reliability of a phylogenetic tree(s) resulting from either current or previous analyses on the basis of the posterior probability distribution of trees approximated by the MCMC method: the fraction of time that a chain visits any particular tree is a valid approximation of the posterior probability of that tree(s). Among the paired-sites tests, the nonparametric ML methods are the most widely used. They include the Kishino and Hasegawa (1989) test (KH test), the Shimodaira and Hasegawa (1999) test (SH test) and its weighted version (WSH test), and the approximately unbiased (AU) test (Shimodaira 2002). The KH test was developed for estimating the standard error and confidence intervals for the difference in log-likelihoods between two phylogenetic trees specified a priori. Shimodaira and Hasegawa (1999) proposed a similar test but making the appropriate allowance for the method to compare a priori and a posteriori topologies and to correct for multiple comparisons. However, Strimmer and Rambaut (2002) pointed out that the SH test may be conservative as the number of trees to be compared increases. This behavior is alleviated in the WSH test (Shimodaira 2002). Finally, Shimodaira (2002) proposed an AU test for assessing the confidence of tree selection that uses a newly devised multiscale bootstrap technique that makes the test less conservative than the SH test (Shimodaira 2002). All these ML topological tests are implemented in PAUP* and CONSEL (Shimodaira and Hasegawa 2001).

10.4.2 Joint Estimation of Alignment and Phylogeny

All commonly accepted methods for phylogenetic reconstruction use as input a single estimate of the alignment that is assumed to be correct. This assumption can lead to exaggerated support for inferred phylogenies if the MSA contains ambiguous regions because near-optimal alignments are ignored (Lutzoni et al. 2000). In addition, the use of progressive algorithms can lead to phylogenies that are biased towards the fixed guide tree assumed in generating the MSA (Redelings and Suchard 2005). However, if the final goal is to generate a phylogenetic tree, there are algorithms for simultaneously (as opposed to sequentially) estimating MSA and trees that relate the sequences within a MP, ML, or Bayesian framework. One such approach is known as direct optimization and is implemented in POY (Wheeler 1996). POY simultaneously estimates ancestral sequences and their pairwise alignment to neighboring sequences by minimizing the number of mutations (substitutions and indels) or maximizing the score under MP and ML optimality criteria, respectively. In both tree searching and character optimization, POY provides the user with complete control over the search, implementing most of the more recently developed algorithms for tree-space searching (e.g., ratchet) and four character optimization algorithms. Recent simulations under the MP criterion have indicated that POY performs worse than Clustal coupled with a subsequent phylogeny search at both estimating alignments and estimating phylogenies (Ogden 2007). Within a Bayesian framework using MCMC techniques, Redelings and Suchard (2005) have proposed a novel evolutionary model and algorithm that can simultaneously estimate and assess confidence in MSA and phylogenies using posterior probabilities. The appeal of this approach is that it allows for the consideration of myriad near-optimal MSAs when estimating phylogenies. These MSAs are weighted by their posterior probabilities, providing objective estimates of uncertainty in the alignment and taking into account information in ambiguous regions. Additionally, this procedure allows for more accurate substitution and indel models of evolution than is possible with sequential methods. This Bayesian method is implemented in the program BALi-Phy (Redelings and Suchard 2005). Naturally, joint estimation of alignments and phylogenies has an associated large cost in computational time, which can preclude the analyses of even medium-sized datasets (approximately 50 taxa).

10.4.3 Networks

When estimating evolutionary relationships among viruses, the reticulating impact of recombination becomes a significant issue. If recombination is present among the sequences of a sample, the evolutionary history among those sequences no longer fits a bifurcating model and therefore a tree representation fails to accurately portray a reasonable genealogy. Under such circumstances, network approaches have been used to represent reticulating genealogical relationships (reviewed in Posada and Crandall 2001). Indeed, such approaches have not only been used to represent reticulate relationships among sequences from a population (e.g., HIV sequences from within a single patient, Wain-Hobson et al. 2003), but might also better represent evolutionary relationships at the origin of life (Rivera and Lake 2004). While there are many different approaches and there is much different software available for estimating reticulate relationships, we are only aware of a single study that actually compares different approaches of network reconstruction. Cassens et al. (2005) compared minimum-spanning network (Excoffier and Smouse 1994) reconstruction via the program Arlequin (Schneider et al. 2000), median-joining networks (Bandelt et al. 1999) implemented in the program Network (http://www.fluxus-engineering.com/sharenet.htm), and statistical parsimony (Templeton et al. 1992) implemented in the program TCS (Clement et al. 2000) with their own algorithm for combining a set of estimated most parsimonious trees into a parsimony network (union of maximum parsimonious trees, UMP). Using simulated sequence evolution without recombination, they found that the UMP method performs well and that UMP, statistical parsimony, and median-joining networks provide better estimates of the true genealogy under broad conditions in terms of sampling of internal nodes, whereas the minimum-spanning network showed very poor performances, especially when internal nodes were poorly sampled. So far, these approaches have not been compared via computer simulation under conditions of recombination where reticulate methods would be expected to outperform bifurcating tree methods.

10.5 Population Inference

Maynard-Smith (1995) pointed out the need for population-genetic insights when contemplating the evolutionary fate of viral pathogens. Population genetics is important in understanding the evolutionary history, epidemiology, and population dynamics of pathogens, the potential for and mode of the evolution of resistance, and ultimately for control strategies. The key factors in the evolutionary response of pathogens to their environments can be measured by assessing the genetic diversity (and partitioning of that diversity within versus between populations), the impact of natural selection in shaping that existing diversity, and the impact of recombination in redistributing that diversity, sometimes into novel combinations. In the previous sections, we described bifurcating and network phylogenetic approaches that can be applied for inferring population structure. The inferred population histories allow us to partition ongoing recurrent evolutionary forces (e.g., gene flow, system of mating) from occasional historical events that impact the demography of the population and the distribution of genetic diversity (e.g., bottlenecks, range expansion, fragmentation). In this section, we describe complementary methods for inferring population demographic history and estimating population parameters.

10.5.1 Inferring Demographic History

Occasionally in the evolutionary history of a species, there are singular demographic events that can leave a lasting impression on the partitioning of population-genetic variation within and among populations (e.g., vicariant events, bottlenecks, founder events, etc.). There are a wide variety of methods for inferring population histories from population-genetic data. These methods vary tremendously in terms of their requirements and assumptions (reviewed in Emerson et al. 2001; Pearse and Crandall 2004). Some methods are based on a supporting phylogeny requiring a molecular clock (Strimmer and Pybus 2001; Drummond et al. 2005), while others require an underlying genealogy but relax the molecular clock assumption and allow for ambiguity in the genealogical estimate (Templeton 1998). Very few account for temporal sampling of viral populations (Drummond et al. 2002; Pybus and Rambaut 2002). Still others avoid evolutionary history altogether (Wooding and Rogers 2002). Yet many argue that there is significant information concerning the population history contained within the genealogy (Epperson 1999; Williamson and Orive 2002), and this can be coupled with other information such as codon usage in protein-coding sequences for a more powerful inference of population history and associated parameter estimates (McVean and Vieira 2001; Drummond et al. 2005). Since these approaches have been extensively reviewed elsewhere, we will not detail them here.

Only a single method, to our knowledge, makes explicit use of both geographical location information as well as genealogical information to allow both spatial and temporal partitioning of historical events and ongoing evolutionary processes, that is, the nested clade phylogeographic analysis (Templeton 1998, 2004). This approach estimates genealogical relationships among sequences using the program TCS (Clement et al. 2000). The resulting genealogy is then used to define a nested hierarchy of genetic relatedness that allows the partitioning of events across relative evolutionary time (i.e., lower nesting levels are more recent events compared with deeper nesting levels; Templeton and Sing 1993; Crandall 1996). Geographic partitioning is accomplished by testing for statistically significant large or small geographic distances among samples relative to their genealogical distance using the program GeoDis (Posada et al. 2000). This allows for the inference of a diverse array of historical population events, including isolation by distance, range expansion, and fragmentation (Templeton 2004).

10.5.2 Inferring Recombination, Genetic Diversity, and Growth

Population parameters of genetic diversity, recombination, and growth can be efficiently estimated using explicit statistical models of evolution such as the coalescent approach, which describes its effect on gene sequences by linking demographic history with population genealogy (Hudson 1991; Nordborg 2001; Felsenstein 2004). Approaches based on this model provide estimates that reflect the evolutionary history of the population rather than the current allele-frequency distribution (Crandall et al. 1999). They use stochastic reduction in lineage number looking backwards through time to infer the past demographic history of the population based on a model of evolution for the marker being used. By their nature, they rely on computationally intense statistical methods and large datasets to make accurate inferences based on genetic data. Nevertheless, considering the speed of personal computers these days, the standardization of sequencing procedures for analyzing large numbers of samples and genes (e.g., multilocus sequence typing, MLST), and the large population sizes available for most microorganisms, we do not think that these are serious limitations for the implementation of coalescent methods to the study of viral population dynamics. Moreover, the coalescent model has several advantages, such as the ease of comparison between genes or species, the ability to make predictions about the question of interest, and the potential to test whether the model of evolution is an adequate characterization of the underlying process (McVean et al. 2002). A more detailed treatment of coalescent theory is beyond the scope of this chapter, but we refer the reader to reviews by Hudson (1991), Nordborg (2001), and Stephens (2001).

10.5.2.1 Recombination

Recombination is generally defined as the exchange of genetic information between two nucleotide sequences. It influences biological evolution at many different levels: it reshuffles existing variation and creates new allele variants, shapes the structure of populations and the action of natural selection, and breaks down linkage disequilibrium (Posada and Crandall 2001). Further, recombination confounds our attempts to infer phylogenetic history (Posada and Crandall 2002) and other key population parameters (Schierup and Hein 2000). Therefore, a clear understanding of how we can detect and estimate the rate at which recombination occurs is essential. A comprehensive review of statistical methods for detecting recombination (test for the occurrence of recombination, identifying the parental and recombinant individuals, and determining the location of breakpoints) and estimating recombination rates in related DNA sequences (i.e., homologous recombination) is presented in Posada et al. (2002) with a complete list of references describing each method and software implementation. The performance of these methods is also reviewed in Posada et al. (2002) and the references therein. Recombination detection methods differ in performance depending on the amount of recombination, the genetic diversity of the data, and the degree of rate variation among sites. As the authors concluded, one should not rely on a single method to detect recombination. No more conclusive are the simulation studies comparing estimators of recombination rates (Wall 2000; Fearnhead and Donnelly 2001). Discrepancies between them are presumably due to the different criteria of assessment and simulation conditions used (Posada et al. 2002).

Many studies of viral population dynamics are only concerned with the detection of recombination, but to understand the role of this force in the generation of genetic diversity we need to accurately estimate the rate at which recombination occurs. Indeed, recombination rate estimators can be used to build tests for the presence of recombination (e.g., likelihood permutation test). They can also be used to indirectly assess the impact of recombination in phylogenetic inference (Pérez-Losada et al. 2006).

10.5.2.2 Genetic Diversity

Genetic diversity (θ) is usually described as $2N_e\mu$ or $4N_e\mu$ in haploid and diploid organisms, respectively. N_e is the effective population size and μ is the mutation rate in mutations per generation. θ can be interpreted as 2 times the neutral mutation rate times the number of heritable gene copies in the population. The units of μ can be mutations per site per generation or mutations per locus per generation. To convert the former into the latter you must multiply the per-site θ by the number of sites at a given locus. If you have information about either population size or mutation rate, for example, mutation rates from molecular biology studies (Mansky and Temin 1995), you can then estimate the other parameter directly. A review of classical and recent statistical methods for estimating genetic diversity is presented in Pearse and Crandall (2004). For the previously discussed reasons we strongly recommend coalescent estimators of θ, such as those implemented in LAMARC (Kuhner et al. 2005) or IM (Hey and Nielsen 2004). However, McVean et al. (2002) describe a corrected version of the classical algorithm of Watterson (1975) for estimating θ that allows for the occurrence of multiple mutations at particular sites (i.e., finite-sites model), and is especially applicable to fast-evolving genomes such as those of some bacteria and viruses. This estimator relies on the number of segregating sites in the sequences and it has been shown that, although less efficient than coalescent ML, it is still remarkably good (Fu and Li 1993; Felsenstein 2004). We recommend its use as an alternative to the more CPU intensive full likelihood approaches.

10.5.2.3 Growth

Another key parameter for characterizing viral population dynamics is the exponential growth rate (g), which shows the relation between θ, now defined as the estimate of modern-day population size, and population size in the past through the equation $\theta_t = \theta_{now}e^{-gt}$, where t is a time in the past. Positive values of g indicate population growth or expansion, negative values indicate population decline, and a zero value indicates that the population has remained constant. Analytical and simulation results have shown that the estimate of g under this model is biased upwards when a finite number of individuals are sampled (Kuhner et al. 1998). Moreover, although we think that the exponential model of growth is particularly suitable for microorganisms, there is typically no a priori reason to make this assumption for a given population. Other methods exist that relax this assumption, such as the skyline plot method of Strimmer and Pybus (2001) implemented in the program GENIE (Pybus and Rambaut 2002), but they also suffer from other problems. The skyline plot, for example, assumes a single evolutionary history (instead of performing an importance sampling scheme as in LAMARC, see below), which should result in less accurate estimates. However, this limitation has recently been overcome by the incorporation of a coalescent Bayesian skyline approach (Drummond et al. 2005) that allows sampling across a set of alternative phylogenies. Such a method is implemented in the program BEAST (Drummond and Rambaut 2003). BEAST includes constant and exponential models of multilocus population growth under different substitution models (including general time reversible and rate heterogeneity). It can also estimate divergence times (t) under constant and local rate molecular clock models and, more interestingly, allows for the analysis of temporally spaced sequence data, such as those collected from populations of rapidly evolving pathogens (e.g., HIV).

Coalescent estimates of recombination, genetic diversity, and exponential growth rates, all together or separately for multiple DNA loci collected from one or multiple populations can be performed in LAMARC. Even if one is simply interested in one of these forces, their simultaneous estimation means that the estimates will not be biased by the unacknowledged presence of another influence. LAMARC duplicates almost exactly the functionality of COALESCE, RECOMBINE, MIGRATE or FLUCTUATE and implements both ML and Bayesian searches of population parameters. The program allows for very refined searches under different models of evolution (including the GTR model), it can accommodate rate heterogeneity (although its implementation is not straightforward), and, importantly, calculates approximate confidence intervals for estimates under the ML search or credibility intervals under the Bayesian search. LAMARC also estimates migration rates (m), although this parameter is not usually of concern for virologists because of the biological characteristics of the organisms under study. However, some interesting studies have been published that trace historical human demographics by looking at migration rates of intestinal pathogens such as *Helicobacter pylori* (Falush et al. 2003), and similar approaches could be used to explore and date the domestication events of various crop species, if the natural hosts were known.

The LDhat package (McVean et al. 2002) estimates population recombination rates (ρ) also within a coalescent framework using the composite likelihood method of Hudson (2001), but adapted to finite-sites models and to estimate variable recombination rates. This method has the desirable property of relaxing the infinite-sites assumption (i.e., mutations only occur once per site in a population) and accommodates different models of molecular evolution (including, importantly, rate heterogeneity). LDhat also includes a powerful likelihood permutation test to test the hypothesis of no recombination ($\rho=0$) as well as other noncoalescent methods for estimating θ and testing the presence of recombination. Finally, LDhat implements the corrected version of the algorithm of Watterson (1975) described in Sec. 10.5.2.2 for estimating θ. Carvajal et al. (2006) have augmented this approach from a two-allele model to a four-allele model and shown it to be robust to a variety of assumption violations common to viral data (rate heterogeneity, population growth, noncontemporaneous sampling, and natural selection).

Multilocus coalescent estimates of θ, m, and t using a MCMC search can also be obtained in IM (Hey and Nielsen 2004). IM applies the isolation with migration model (Hey and Nielsen 2004) to genetic data drawn from a pair of closely related populations or species. The results are estimates of the marginal posterior probability densities for each of the population parameters under study. The program implements four mutation models and assumes no recombination within loci.

10.5.3 Inferring Adaptive Evolution

The importance of selection in molecular evolution is still a matter of debate. The neutral theory (Kimura 1983) maintains that most observed molecular variation is due to random fixation of selectively neutral mutations. Many studies, however, have detected adaptive selection (i.e., Darwinian selection fixing advantageous mutations with positive selective coefficients) in protein-coding genes from diverse organisms, and a vast number of those involve microbial organisms. A few good examples of those include genes involved in defensive systems, drug resistance, evading the immune system, ATP synthesis, and DNA replication (Yang and Bielawski 2000; Anisimova et al. 2003).

When studying adaptive selection one must distinguish between the two different inferential problems of testing for positive selection in a particular gene or section of a gene and of predicting which sites are most likely to be under positive selection. The methods described next attempt to address these two questions independently. To our knowledge no recent comprehensive review of these methods has been published, although many studies on this topic exist. We will take this opportunity to compile them here in a synthetic review.

10.5.3.1 Evaluating Positive Selection in Terms of d_N/d_S Ratios

The standard method for detecting adaptive molecular evolution in protein-coding DNA sequences is through comparison of nonsynonymous (AA-changing; d_N) and synonymous (silent; d_S) substitution rates through the d_N/d_S ratio (ω or acceptance rate; Miyata and Yasunaga 1980). ω measures the difference between both rates on the basis of a codon-substitution model. If an AA substitution is neutral, it will be fixed at the same rate as a synonymous mutation, with $\omega = 1$. If the AA change is deleterious, purifying or negative selection (i.e., natural selection against deleterious mutations with negative selection coefficients) will reduce its fixation rate; thus, $\omega < 1$. Only when the AA change offers a selective advantage is it fixed at a higher rate than a synonymous mutation, with $\omega > 1$. Therefore, an ω ratio significantly higher than 1 is convincing evidence for adaptive or diversifying selection. Basically, three classes of methods have been proposed for detecting if a protein is experiencing an excess of nonsynonymous substitutions or elevated values of ω: approximate or ad hoc methods, MP, and ML methods:

1. *Approximate or ad hoc methods*. Since the early 1980s several intuitive methods have been proposed to estimate averaged (gene-specific) ω. These methods make simplistic assumptions about the nucleotide substitution and involve ad hoc treatments that cannot be justified rigorously. Among them, the most commonly used and the one preferred by many virologists is the method of Nei and Gojobori (1986), which is implemented in the program MEGA (Kumar et al. 2004). This method relies on the JC69 (Jukes and Cantor 1969) nucleotide substitution model, ignores the transition/transversion rate bias, and does not include a codon model that accounts for the codon-usage bias (i.e., unequal codon frequencies in a gene). Computer simulations and analytical analyses have demonstrated that ignoring these factors leads to inaccurate estimates of the ω ratio (Yang and Bielawski 2000). More recent ad hoc methods, however, have been proposed that account for these biases and include more complex models of DNA substitution (Yang and Nielsen 2000), although these approaches are less powerful than those based on site-specific models of adaptive selection (see later).
2. *Maximum parsimony estimation*. Parsimony methods were independently developed by Fitch et al. (1997) and Suzuki and Gojobori (1999). In these methods, substitutions are inferred using parsimony reconstruction of ancestral sequences, and an excess of nonsynonymous substitutions is tested independently for each site. Under these methods, in order to detect positive selection in a gene where multiple sites are analyzed, a correction for multiple testing (e.g., Bonferroni or its improved version by Simes 1986) is needed. The Suzuki and Gojobori (1999) method (more popular) is implemented in the computer program ADAPTSITE of Suzuki et al. (2001). ADAPTSITE also includes a distance-based Bayesian method (Zhang and Nei 1997) for inferring ancestral codons.
3. *ML estimation*. ML methods are based on explicit models of codon substitution (Goldman and Yang 1994). Models include parameters such as branch lengths, codon frequencies, and transition to transversion rate ratio, which are estimated

from the data (i.e., they account for possible biases). Thus, estimates of ω from ML are expected to be more reliable than those generated from previous approximate or parsimony methods (Yang and Nielsen 2000). Nevertheless, approximate and former ML methods such as that of Goldman and Yang (1994) calculate the ω ratio as an average over all codon sites in the gene and over the entire evolutionary time that separates the sequences (i.e., all lineages in the phylogeny). The criterion that this average ω be greater than 1 is a very stringent one for detecting adaptive selection (Crandall et al. 1999). Most variation within genes that encode essential metabolic enzymes, such as the MLST housekeeping genes, is considered neutral or deleterious owing to functional constraints (Li 1997; Feil et al. 2000, 2003; Dingle et al. 2001; Meats et al. 2003; Urwin and Maiden 2003). Adaptive evolution most likely occurs at a few time points and affects a few AAs. Therefore, in such cases, the ω averaged over time and over sites will not be significantly greater than 1 even if adaptive molecular evolution has occurred. But ML is a powerful and flexible method for estimating parameters and testing hypotheses, so complex evolutionary scenarios can be devised within statistical models. Nielsen and Yang (1998) and Yang et al. (2000) implemented 13 new evolutionary models (statistical distributions) that build on the ML model of Goldman and Yang (1994) but allow for heterogeneous ω ratios among sites in a phylogeny (i.e., they do not account for variation of ω among lineages). Among them, the authors recommended the use of M1 (neutral), M2 (selection), M3 (discrete), M7 (beta), and M8 (beta&ω) (see Table 2 in Yang et al. 2000 for details). Models M1 and M7 do not allow for positively selected sites (with $\omega > 1$), but models M2, M3, and M8 add extra parameters mainly to account for the possible occurrence of positive selection. The log-likelihood under a model measures the fit of the model to the data, and we can compare two models by comparing their log-likelihood values (LRT). Yang et al. (2000), Yang and Nielsen (2002), and Anisimova et al. (2003) noticed that the M0 versus M3 comparison is really a test of variability of selective pressures among sites (so it does not constitute a rigorous test of positive selection), whereas the M1 versus M2 or M3 and M7 versus M8 comparisons are tests of positive selection. The good performance of these site-specific models is well documented (Anisimova et al. 2003; Pérez-Losada et al. 2005). Results of a more extensive study based on 91 MLST loci (presumably neutral) corresponding to one fungal and 16 bacterial pathogens can be found in Pérez-Losada et al. (2006).

The previous models tested under the LRT are still conservative, as they require that positively selected sites be under diversifying selection along all lineages on the phylogeny. Hence, the ML model of codon substitution of Goldman and Yang (1994) has also been extended to account for variation of ω among lineages and sites (Yang and Nielsen 2002). It appears that averaging over sites is a more serious problem than averaging over lineages because the site-specific analysis has been successful in detecting positive selection in a variety of genes (Yang et al. 2000; Pérez-Losada et al. 2005, 2006). Computer simulations also confirmed the power of the site-specific analysis (Yang and Bielawski 2000; Anisimova et al. 2001).

Yang and Nielsen (2002) implemented two new versions of their site- and lineage-specific models that are useful for identifying positive selection along prespecified lineages that affect only a few sites in the protein. Recently, Guindon et al. (2004) introduced two new models that allow selection to change over time, but unlike previous methods their approach does not constrain switches among selection categories to any particular lineage a priori. The program that implements these models is available at http://www.cebl.sbs.auckland.ac.nz/stephane/fitmodel.html.

The abovementioned site- and/or lineage-specific ML models assume that there are several heterogeneous site classes but we do not know a priori which class each site is from. Those models are referred to as random-site models (Yang and Swanson 2002). Sometimes prior information is available to partition sites into classes, which are expected to have different selective pressures and thus different ω ratios (e.g., combined analysis of C and V domains from the HIV *env* gene). In such cases, it is reasonable to make use of such information and fit models that assign different ω ratios for site classes. Models that account for the heterogeneity of different site partitions (fixed-site models) are implemented in Yang and Swanson (2002).

The last ML approach we include in this section was independently developed by Forsberg and Christiansen (2003) and Bielawski and Yang (2004). They presented a site-specific ML method useful for measuring divergence selective pressures between clades, such as between new and original host species clades (host radiation) in a parasite (Forsberg and Christiansen 2003) or between paralogous clades of a gene family (gene duplication) (Bielawski and Yang 2004). The codon-substitution models developed by both groups are similar and build on previous ML methods reported by Goldman and Yang (1994), Yang et al. (2000), and Yang and Nielsen (2002). The utility of these methods is illustrated on datasets of nucleoprotein sequences from the influenza A virus obtained from avian and human hosts and two presented examples of gene duplication: the ε and γ globins and two eosinophil proteins.

Until now we have shown different uses of the LRT to test for positive selection in a gene as a whole under different models of codon change (first step). If this test indicates statistical evidence for the presence of sites evolving under positive selection, identification of those sites would be desirable (second step). Nielsen and Yang (1998) proposed an empirical Bayes approach (known as the naive empirical Bayes, NEB) for predicting which sites are most likely to be under positive selection. NEB is used to calculate the posterior probability that each site is from a particular site class, and sites with high posterior probability (say, 0.95 or higher) coming from the class with $\omega > 1$ are inferred to be under positive selection. NEB uses ML estimates of parameters, such as the ω ratios for the site classes, without accounting for their sampling errors, and the NEB calculation of posterior probability may be unreliable in small datasets lacking phylogenetic signal (Anisimova et al. 2002). Yang et al. (2005) developed a new Bayes empirical Bayes (BEB) method that accommodates uncertainties in the ML estimates of parameters in the ω distribution using numerical integration. The authors tested the method in real data and using computer simulations, and showed that BEB in small datasets does not generate false positives as did the NEB approach, while in large datasets it retains the power of the NEB approach for inferring positively selected sites. Both NEB and BEB methods have been implemented in previously described site-, lineage- and clade-specific ML models.

The diversity of ML methods and tests described above may look intimidating in terms of software implementation. Fortunately all of them but the model of Forsberg and Christiansen (2003) (http://birc.dk) are carried out by the software package PAML of Yang (1997) under different variants of the program codeml. A manual and examples explain in detail how to perform those analyses.

High recombination rates can affect both the estimation of d_N/d_S and sites under adaptive selection (see below). A new coalescent model was recently described that estimates the d_N/d_S ratio and the population recombination rate (ρ) simultaneously in omegaMap (Wilson and McVean 2006) and can be run under a constant (i.e., homogeneous ω and ρ) and independent (i.e., heterogeneous ω and/or ρ) models.

Another excellent program that offers a variety of tests for detecting selection, and also includes a tutorial with many examples, is HyPhy (Kosakovsky Pond et al. 2005). This program includes a versatile suite of methods to detect adaptive evolution at individual AA sites and/or lineages, including generalizations of PAML and ADAPTSITE approaches and many others.

10.5.3.2 Evaluating Positive Selection in Terms of Amino Acid Properties

McClellan et al. (2005) have shown recently using conservative cytochrome *b* sequences that d_N/d_S ratios are less sensitive to detecting single adaptive AA changes than methods that evaluate positive selection in terms of the AA properties that comprise proteins. They estimated adaptive selection in terms of 31 quantitative biochemical properties using the computer program TreeSAAP (Woolley et al. 2003). Based on a phylogenetic tree, a chronology of observable molecular evolutionary events using the evolutionary DNA models implemented in the PAML algorithm baseml is first established. TreeSAAP then compares sequences in the context of the specified phylogenetic topology, codon by codon, to infer AA replacement events. The inferred pattern of AA replacement is then analyzed by using the models of Xia and Li (1998) and McClellan and McCracken (2001). Both models estimate distributions of potential changes in physicochemical AA properties by assuming that every possible AA replacement is equally likely under neutral conditions. Expected and observed mean changes in AA properties and the relative shapes of expected and observed distributions are finally compared using different basic statistics in order to identify (1) AA properties that may have changed more or less often than expected by chance and (2) AA sites associated with selection, thus establishing a correlation between the sites under selection and the structure and function of the protein. This approach is broader in scope than estimating d_N/d_S ratios since it allows for testing not only the presence (positive) or absence (negative) of adaptive selection (referred to as destabilizing selection: selection that results in radical structural or functional shifts in local regions of the protein) but also negative and positive stabilizing selection (selection that results in radical structural or functional constraints). TreeSAAP can assess these two types of selection over the entire dataset or by regions or domains (window analyses) specified by the user, with the latter being a more sensitive analysis. Examples implementing the TreeSAAP approach can be found in McClellan et al. (2005), Pérez-Losada et al. (2005,2006), and Taylor et al. (2005).

There are also ML methods that assess selection in terms of AA properties (Sainudiin et al. 2005). These methods build on the codon-based models of Nielsen and Yang (1998) and Yang et al. (2000) to provide a likelihood framework to detect an elevation in the rate of property-altering to property-conserving substitutions. The NEB method of Nielsen and Yang (1998) is used to compute the posterior probability that a particular site is subject to an elevated rate ratio (greater than 1). The method can be applied to any physicochemical property of interest by partitioning the codons according to that property instead of partitioning the codons on the basis of the encoded AAs. These models are implemented in PAML.

10.5.3.3 Limitations and Practical Considerations

We already addressed the serious caveats affecting approximate methods for detecting selection, so we do not strongly recommend their use. ML and MP methods rely on the phylogenetic relationship among the sequences, so one should provide the best tree possible. PAML can generate a tree, but the tree reconstruction algorithm included is not very efficient, so its use is not recommended. Nevertheless, analyses by Yang et al. (2000) and Ford (2001) suggest that the LRT and the Bayes inference of sites under selection do not seem to be sensitive to the assumed topology. We do not know how this factor may affect the other methods.

Methods based on ancestral reconstruction (parsimony and TreeSAAP methods) might not provide reliable statistical tests because they ignore errors and biases in reconstructed ancestral sequences (although this problem is alleviated under the Bayesian approach) and involve systematic biases (the site-class models also suffer from this problem) (Yang and Bielawski 2000; Anisimova et al. 2001). An evaluation of the Suzuki and Gojobori (1999) parsimony method as implemented in ADAPTSITE has been given by Wong et al. (2004). The authors concluded that the lack of power of this method makes it unusable for testing positive selection except in large datasets with many sequences, which agrees with Suzuki and Gojobori's original study; hence, ADAPTSITE should be used only for exploratory analyses. More extensive analyses by Pond and Frost (2005) also showed that this method is only suitable for large alignments.

Methods based on comparisons of d_N and d_S rates also have limitations. Such methods only detect positive selection if there is an excess of nonsynonymous substitutions and they are thus suitable for detecting recurrent diversifying selection, but may not detect directional selection that drives an advantageous mutation quickly to fixation. These methods assume that the sites or lineages have a constant selective pressure over evolutionary time. They may be powerful at detecting

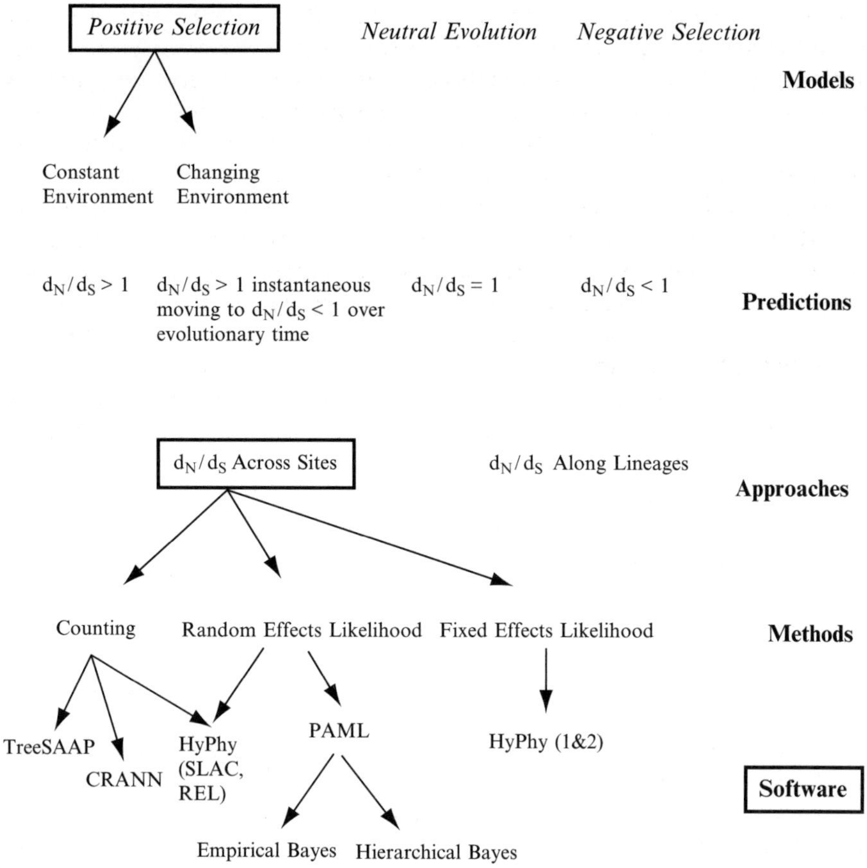

Fig. 10.2 Summary of some available methods for detecting positive selection using d_N/d_S ratios for protein-coding sequences

evidence of selection when the selective force requires multiple genetic changes over time (e.g., changes in viral evolution in response to immune pressure – the changing environment model). But these methods may not identify selected sites that confer a fitness advantage and then become fixed in the population (e.g., variants associated with resistance – the positive selection model) (Fig. 10.2). Thus, a reasonable amount of d_N and d_S is necessary for such methods to work, as too little information is available at low divergence levels, while synonymous substitutions are often saturated at high divergence. Simulations reported by Anisimova et al. (2001, 2002, 2003) and Shriner et al. (2003) assessed the accuracy and power of the LRT and Bayes test under different conditions of sequence divergence, sequence length, number of taxa, strength of positive selection, and recombination. General conclusions from these analyses indicate that the LRT is conservative, especially when the data contain very short and highly similar sequences and fewer than approximately 15 taxa. In small datasets the BEB does not generate false positives as did the old NEB approach. Excessive recombination ($\rho = 0.01$), often observed in some virus (e.g., HIV) and bacteria (e.g., *Helicobacter pylori*) populations can also cause false positives and make the LRT unrealistic as it often mistakes recombination as evidence for positive selection. The LRT test that compares models M7 and M8 seems to be more robust to recombination. The detection of sites under positive selection seems to be less affected by recombination; however, using methods that explicitly account for recombination (e.g., omegaMap) when inferring both selection rates and sites seems more appropriate. We encourage the reader to review those studies for more details.

Unfortunately, at the moment, methods for detecting selection are focused on positive diversifying selection on protein-coding regions in a constant environment. We are unaware of methods for detecting selection in RNA based on structural models. Future approaches will hopefully incorporate the reality of changing environments and non-protein-coding sequences as it is clear that natural selection is acting on these sequences as well.

10.6 Summary

The fields of phylogenetics and population genetics offer a broad array of tools for the sophisticated analyses of plant viral sequence data. Both fields are rapidly evolving with new and better methods being developed with every issue of associated journals. It is nearly impossible to keep up with new developments in these fields as well as in specific areas of virology. We therefore recommend collaborations between virologists and evolutionary biologists to reap the most out of the truly rich datasets being collected and available today (Tibayrenc 2005).

Acknowledgements We thank Marilyn Roossinck for her cordial invitation to write this chapter and her seemingly infinite patience. This work was supported by NIH grants R01 AI50217 and GM66276 and Brigham Young University.

References

Abascal F, Zardoya R, Posada D (2005) ProtTest: selection of best-fit models of protein evolution. Bioinformatics 21:2104–2105

Abdo Z, Minin VN, Joyce P, Sullivan J (2005) Accounting for uncertainty in the tree topology has little effect on the decision-theory approach to model selection in phylogeny estimation. Mol Biol Evol 22:691–703

Adachi J, Hasegawa M (1996) MOLPHY version 2.3: programs for molecular phylogenetics based in maximum likelihood. Comput Sci Monogr 28:1–150

Adachi J, Waddell PJ, Martin W, Hasegawa M (2000) Plastid genome phylogeny and a model of amino acid substitution for proteins encoded by chloroplast DNA. J Mol Evol 50:348–358

Alfaro ME, Zoller S, Lutzoni F (2003) Bayes or bootstrap? A simulation study comparing the performance of Bayesian Markov chain Monte Carlo sampling and bootstrapping in assessing phylogenetic confidence. Mol Biol Evol 20:255–266

Anisimova M, Bielawski JP, Yang Z (2001) Accuracy and power of the likelihood ratio test in detecting adaptive molecular evolution. Mol Biol Evol 18,1585–1592

Anisimova M, Bielawski JP, Yang Z (2002) Accuracy and power of Bayes prediction of amino acid sites under positive selection. Mol Biol Evol 19:950–958

Anisimova M, Nielsen R, Yang Z (2003) Effect of recombination on the accuracy of the likelihood method for detecting positive selection at amino acid sites. Genetics 164:1229–1236

Aris-Brosou S, Yang Z (2002) Effects of models of rate evolution on estimation of divergence dates with special reference to the metazoan 18S ribosomal RNA phylogeny. Syst Biol 51:703–714

Bandelt HJ, Forster P, Rohl A (1999) Median-joining networks for inferring intraspecific phylogenies. Mol Biol Evol 16:37–48

Bielawski JP, Yang Z (2004) A maximum likelihood method for detecting functional divergence at individual codon sites, with application to gene family evolution. J Mol Evol 59:121–132

Brauer MJ, Holder MT, Dries LA, Zwickl DJ, Lewis PO, Hillis DM (2002) Genetic algorithms and parallel processing in maximum-likelihood phylogeny inference. Mol Biol Evol 19:1717–1726

Bremer K (1988) The limits of amino acid sequence data in angiosperm phylogenetic reconstruction. Evolution 42:795–803

Browne M (2000) Cross-validation methods. J Math Psychol 44:108–132

Bruno WJ, Halpern AL (1999) Topological bias and inconsistency of maximum likelihood using wrong models. Mol Biol Evol 16:564–566

Buckley TR (2002) Model misspecification and probabilistic tests of topology: evidence from empirical data sets. Syst Biol 51:509–523

Buckley TR, Cunningham CW (2002) The effects of nucleotide substitution model assumptions on estimates of nonparametric bootstrap support. Mol Biol Evol 19:394–405

Burnham KP, Anderson DA (2002) Model selection and multimodel inference: a practical information-theoretic approach, 2nd edn. Springer, New York

Cao Y, Janke A, Waddell PJ, Westerman M, Takenaka O, Murata S, Okada N, Paabo S, Hasegawa M (1998) Conflict among individual mitochondrial proteins in resolving the phylogeny of eutherian orders. J Mol Evol 47:307–322

Carvajal-Rodríguez A, Crandall KA, Posada D (2006) Recombination estimation under complex evolutionary models with the coalescent composite likelihood method. Mol Biol Evol 13:817–827

Cassens I, Mardulyn P, Milinkovitch MC (2005) Evaluating intraspecific "network" construction methods using simulated sequence data: do existing algorithms outperform the global maximum parsimony approach? Syst Biol 54:363–372

Castresana J (2000) Selection of conserved blocks from multiple alignments for their use in phylogenetic analysis. Mol Biol Evol 17:540–552

Clement M, Posada D, Crandall KA (2000) TCS: a computer program to estimate gene genealogies. Mol Ecol 9:1657–1659

Crandall KA (1996) Multiple interspecies transmissions of human and simian T-cell leukemia/lymphoma virus type I sequences. Mol Biol Evol 13:115–131

Crandall KA, Kelsey CR, Imamichi H, Lane HC, Salzman NP (1999) Parallel evolution of drug resistance in HIV: failure of nonsynonymous/synonymous substitution rate ratio to detect selection. Mol Biol Evol 16:372–382

Crandall KA, Pérez-Losada M, Christensen RG, McClellan DA, Viscidi RP (2005) Phylogenomics and molecular evolution of polyomaviruses. In: Ahsan N (ed) Polyomavirus and human diseases. Landes BioScience, Georgetown, pp 1–14

Dayhoff MO, Schwartz RM, Orcutt BC (1978) A model of evolutionary change in proteins. In: Dayhoff MO (ed) Atlas of protein sequence and structure. National Biomedical Research Foundation, Washington, pp 345–352

Desper R, Gascuel O (2002) Fast and accurate phylogeny reconstruction algorithms based on the minimum-evolution principle. J Comput Biol 9:687–705

Dimmic MW, Rest JS, Mindell DP, Goldstein RA (2002) rtREV: an amino acid substitution matrix for inference of retrovirus and reverse transcriptase phylogeny. J Mol Evol 55:65–73

Dingle KE, Colles FM, Wareing DR, Ure R, Fox AJ, Bolton FE, Bootsma HJ, Willems RJ, Urwin R, Maiden MC (2001) Multilocus sequence typing system for *Campylobacter jejuni*. J Clin Microbiol 39:14–23

Donoghue MJ, Olmstead RG, Smith JF, Palmer JD (1992) Phylogenetic relationships of dipsacales based on rbcL sequences. Ann Mo Bot Gard 79:333–345

Douady CJ, Delsuc F, Boucher Y, Doolittle WF, Douzery EJ (2003) Comparison of Bayesian and maximum likelihood bootstrap measures of phylogenetic reliability. Mol Biol Evol 20:248–254

Drummond AJ, Nicholls GK, Rodrigo AG, Solomon W (2002) Estimating mutation parameters, population history and genealogy simultaneously from temporally spaced sequence data. Genetics 161:1307–1320

Drummond AJ, Rambaut A (2003) BEAST version 1.0. http://evolve.zoo.ox.ac.uk/beast/

Drummond AJ, Rambaut A, Shapiro B, Pybus OG (2005) Bayesian coalescent inference of past population dynamics from molecular sequences. Mol Biol Evol 22:1185–1192

Duret L, Abdeddaim S (2000) Multiple alignment for structural, functional, or phylogenetic analyses of homologous sequences. In: Higgins D, Taylor W (eds) Bioinformatics, sequence, structure and databanks. Oxford University Press, Oxford, pp 155–167

Edgar RC (2004) MUSCLE: multiple sequence alignment with high accuracy and high throughput. Nucleic Acids Res 32:1792–1797

Emerson BC, Paradis E, Thébaud C (2001) Revealing the demographic histories of species using DNA sequences. Trends Ecol Evol 16:707–716

Epperson BK (1999) Gene genealogies in geographically structured populations. Genetics 152:797–806

Excoffier L, Smouse PE (1994) Using allele frequencies and geographic subdivision to reconstruct gene trees within a species: molecular variance parsimony. Genetics 136:343–359

Falush D, Wirth T, Linz B, Pritchard JK, Stephens M, Kidd M, Blaser MJ, Graham DY, Vacher S, Pérez-Pérez GI, Yamaoka Y, Megraud F, Otto K, Reichard U, Katzowitsch E, Wang X, Achtman M, Suerbaum S (2003) Traces of human migrations in *Helicobacter pylori* populations. Science 299:1582–1585

Fearnhead P, Donnelly P (2001) Estimating recombination rates from population genetic data. Genetics 159:1299–1318

Feil EJ, Smith JM, Enright MC, Spratt BG (2000) Estimating recombinational parameters in *Streptococcus pneumoniae* from multilocus sequence typing data. Genetics 154:1439–1450

Feil EJ, Cooper JE, Grundmann H, Robinson DA, Enright MC, Berendt T, Peacock SJ, Smith JM, Murphy M, Spratt BG, Moore CE, Day NP (2003) How clonal is *Staphylococcus aureus*? J Bacteriol 185:3307–3316

Felsenstein J (1978) Cases in which parsimony or compatibility methods will be positively misleading. Syst Zool 27:401–410

Felsenstein J (1981) Evolutionary trees from DNA sequences: a maximum likelihood approach. J Mol Evol 17:368–376

Felsenstein J (1985) Confidence limits on phylogenies: an approach using the bootstrap. Evolution 39:783–791

Felsenstein J (2004) Inferring phylogenies. Sinauer, Sunderland

Fitch WM, Bush RM, Bender CA, Cox NJ (1997) Long term trends in the evolution of H(3) HA1 human influenza type A. Proc Natl Acad Sci. USA 94:7712–7718

Ford MJ (2001) Molecular evolution of transferrin: evidence for positive selection in salmonids. Mol Biol Evol 18:639–647

Forsberg R, Christiansen FB (2003) A codon-based model of host-specific selection in parasites, with an application to the influenza A virus. Mol Biol Evol 20:1252–1259

Frati F, Simon C, Sullivan J, Swofford DL (1997) Evolution of the mitochondrial COII gene in Collembola. J Mol Evol 44:145–158

Fu YX, Li WH (1993) Maximum likelihood estimation of population parameters. Genetics 134:1261–1270

Gatesy J, DeSalle R, Wheeler W (1993) Alignment-ambiguous nucleotide sites and the exclusion of systematic data. Mol Phylogenet Evol 2:152–157

Gaut BS, Lewis PO (1995) Success of maximum likelihood phylogeny inference in the four-taxon case. Mol Biol Evol 12:152–162

Goldman N, Yang Z (1994) A codon-based model of nucleotide substitution for protein-coding DNA sequences. Mol Biol Evol 11:725–736

Graur D, Martin W (2004) Reading the entrails of chickens: molecular timescales of evolution and the illusion of precision. Trends Genet 20:80–86

Green PJ (1995) Reversible jump Markov chain Monte Carlo computation and Bayesian model determination. Biometrika 82:711–732

Guindon S, Gascuel O (2003) A simple, fast, and accurate algorithm to estimate large phyhlogenies by maximum likelihood. Syst Biol 52:696–704

Guindon S, Rodrigo AG, Dyer KA, Huelsenbeck JP (2004) Modeling the site-specific variation of selection patterns along lineages. Proc Natl Acad Sci USA 101:12957–12962

Hendy MD, Penny D (1982) Branch and bound algorithms to determine minimal evolutionary trees. Math Biosci 59:277–290

Henikoff S, Henikoff JG (1992) Amino acid substitution matrices for protein blocks. Proc Natl Acad Sci USA 89:10915–10919

Hey J, Nielsen R (2004) Multilocus methods for estimating population sizes, migration rates and divergence time, with applications to the divergence of *Drosophila pseudoobscura* and *D. persimilis*. Genetics 167:747–760

Hillis DM, Bull JJ (1993) An empirical test of bootstrapping as a method for assessing confidence in phylogenetic analysis. Syst Biol 42:182–192

Hudson RR (1991) Gene genealogies and the coalescent process. In: Futuyma D, Antonovics J (eds) Oxford surveys in evolutionary biology, vol 7. Oxford University Press, New York, pp 1–44

Hudson RR (2001) Two-locus sampling distributions and their application. Genetics 159:1805–1817

Huelsenbeck JP (1995) Performance of phylogenetic methods in simulation. Syst Biol 44:17–48

Huelsenbeck JP, Crandall KA (1997) Phylogeny estimation and hypothesis testing using maximum likelihood. Annu Rev Ecol Syst 28:437–466

Huelsenbeck JP, Hillis DM (1993) Success of phylogenetic methods in the four-taxon case. Syst Biol 42:247–264

Huelsenbeck JP, Ronquist F (2001) MRBAYES: Bayesian inference of phylogeny. Bioinformatics 17:754–755

Huelsenbeck JP, Ronquist F, Nielsen R, Bollback JP (2001) Bayesian inference of phylogeny and its impact on evolutionary biology. Science 294:2310–2314

Huelsenbeck JP, Larget B, Miller RE, Ronquist F (2002) Potential applications and pitfalls of Bayesian inference of phylogeny. Syst Biol 51:673–688

Huelsenbeck JP, Larget B, Alfaro ME (2004) Bayesian phylogenetic model selection using reversible jump Markov chain Monte Carlo. Mol Biol Evol 21:1123–1133

Huson DH, Steel M (2004) Distances that perfectly mislead. Syst Biol 53:327–332

Janies DA, Wheeler WC (2001) Efficiency of parallel direct optimization. Cladistics 17: S71–S82

Jobb G (2005) TREEFINDER version of June 2005. http://www.treefinder.de

Jones DT, Taylor WR, Thornton JM (1992) The rapid generation of mutation data matrices from protein sequences. Comput Appl Biosci 8:275–282

Jukes TH, Cantor CR (1969) Evolution of protein molecules. In: Munro HM (ed) Mammalian protein metabolism. Academic, New York, pp 21–132

Kass RE, Raftery AE (1995) Bayes factors and model uncertainty. J Am Stat Assoc 90:773–775

Katoh K, Kuma K, Miyata T (2001) Genetic algorithm-based maximum-likelihood analysis for molecular phylogeny. J Mol Evol 53:477–484

Katoh K, Kuma K, Toh H, Miyata T (2005) MAFFT version 5: improvement in accuracy of multiple sequence alignment. Nucleic Acids Res. 33:511–518

Katoh K, Misawa K, Kuma K, Miyata T (2002) MAFFT: a novel method for rapid multiple sequence alignment based on fast Fourier transform. Nucleic Acids Res 30:3059–3066

Kimura M (1983) The neutral theory of molecular evolution. Cambridge University Press, Cambridge

Kishino H, Hasegawa M (1989) Evaluation of the maximum likelihood estimate of the evolutionary tree topologies from DNA sequence data, and the branching order in Hominoidea. J Mol Evol 29:170–179

Kosakovsky Pond SL, Frost SDW, Muse SV (2005) HyPhy: hypothesis testing using phylogenies. Bioinformatics 21:676–679

Kuhner MK, Yamato J, Felsenstein J (1998) Maximum likelihood estimation of population growth rates based on the coalescent. Genetics 149:429–434

Kuhner MK, Yamato J, Beerli P, Smith LP, Rynes E, Walkup E, Li C, Sloan J, Colacurcio P, Felsenstein J (2005) LAMARC version 2.0.2. University of Washington

Kumar S, Tamura K, Nei M (2004) MEGA3: Integrated software for molecular evolutionary genetics analysis and sequence alignment. Brief Bioinform 5:150–163

Leaché AD, Reeder TW (2002) Molecular systematics of the Eastern Fence Lizard (*Sceloporus undulatus*): a comparison of parsimony, likelihood, and Bayesian approaches. Syst Biol 51:44–68

Lemmon AR, Milinkovitch MC (2002) The metapopulation genetic algorithm: An efficient solution for the problem of large phylogeny estimation. Proc Natl Acad Sci USA 99:10516–10521

Lewis PO (1998) A genetic algorithm for maximum-likelihood phylogeny inference using nucleotide sequence data. Mol Biol Evol 15,277–283

Li WH (1997) Molecular evolution. Sinauer, Sunderland

Löytynoja A, Milinkovitch MC (2003) A hidden Markov model for progressive multiple alignment. Bioinformatics 19:1505–1513

Lunter G, Miklos I, Drummond A, Jensen JL, Hein J (2005) Bayesian coestimation of phylogeny and sequence alignment. BMC Bioinformatics 6:83

Lutzoni F, Wagner P, Reeb V, Zoller S (2000) Integrating ambiguously aligned regions of DNA sequences in phylogenetic analyses without violating positional homology. Syst Biol 49:628–651

Maddison DR, Maddison WP (2000) MacClade 4: analysis of phylogeny and character evolution, version 4.0. Sinauer, Sunderland

Mansky LM, Temin HM (1995) Lower in vivo mutation rate of human immunodeficiency virus type 1 than that predicted from the fidelity of purified reverse transcriptase. J Virol 69:5087–5094

Matsuda H (1996) Protein phylogenetic inference using maximum likelihood with a genetic algorithm. Paper presented at the Pacific symposium on biocomputing

Maynard Smith J (1995) Do bacteria have population genetics? In: Baumberg JP (ed) Population genetics of bacteria. Press Syndicate of the University of Cambridge, Cambridge, pp 1–12

McClellan DA, McCracken KG (2001) Estimating the influence of selection on the variable amino acid sites of the cytochrome b protein functional domains. Mol Biol Evol 18:917–925

McClellan DA, Palfreyman EJ, Smith MJ, Moss JL, Christensen RG, Sailsbery JK (2005) Physicochemical evolution and molecular adaptation of the cetacean and artiodactyl cytochrome b proteins. Mol Biol Evol 22:437–455

McVean G, Awadalla P, Fearnhead P (2002) A coalescent-based method for detecting and estimating recombination from gene sequences. Genetics 160:1231–1241

McVean GAT, Vieira J (2001) Inferring parameters of mutation, selection and demography from patterns of synonymous site evolution in *Drosophila*. Genetics 157:245–257
Meats E, Feil EJ, Stringer S, Cody AJ, Goldstein R, Kroll JS, Popovic T, Spratt BG (2003) Characterization of encapsulated and noncapsulated *Haemophilus influenzae* and determination of phylogenetic relationships by multilocus sequence typing. J Clin Microbiol 41:1623–1636
Minin V, Abdo Z, Joyce P, Sullivan J (2003) Performance-based selection of likelihood models for phylogeny estimation. Syst Biol 52:674–683
Miyata T, Yasunaga T (1980) Molecular evolution of mRNA: a method for estimating evolutionary rates of synonymous and amino acid substitutions from homologous nucleotide sequences and its application. J Mol Evol 16:23–36
Muller T, Vingron M (2000) Modeling amino acid replacement. J Comput Biol 7:761–776
Nei M, Gojobori T (1986) Simple methods for estimating the numbers of synonymous and nonsynonymous nucleotide substitutions. Mol Biol Evol 3:418–426
Nielsen R, Yang Z (1998) Likelihood models for detecting positively selected amino acid sites and applications to the HIV-1 envelope gene. Genetics 148:929–936
Nixon KC (1999) The parsimony ratchet, a new method for rapid parsimony analyses. Cladistics 15:407–414
Nordborg M (2001) Coalescent theory. In: Balding D, Bishop M, Cannings C (eds) Handbook of statistical genetics. Wiley, Chichester, pp 179–212
Notredame C (2002) Recent progress in multiple sequence alignment: a survey. Pharmacogenetics 3:131–144
Notredame C, Higgins DG, Heringa J (2000) T-Coffee: a novel method for fast and accurate multiple sequence alignment. J Mol Biol 302:205–217
Nylander JAA, Ronquist F, Huelsenbeck JP, Nieves-Aldrey JL (2004) Bayesian phylogenetic analysis of combined data. Syst Biol 53:47–67
Ogden TH, Rosenberg MS (2007) Alignment and topological accuracy of the direct optimization approach via POY and traditional phylogenetics via ClustalW + PAUP*. Syst Biol 56:182–193
O'Sullivan O, Suhre K, Abergel C, Higgins DG, Notredame C (2004) 3DCoffee: combining protein sequences and structures within multiple sequence alignments. J Mol Biol 340:385–395
Page RDM, Holmes EC (1998) Molecular evolution: a phylogenetic approach. Blackwell, Cambridge
Pagel M, Meade A (2004) A phylogenetic mixture model for detecting pattern-heterogeneity in gene sequence of character-state data. Syst Biol 53:571–581
Pearse DE, Crandall KA (2004) Beyond Fst: analysis of population genetic data for conservation. Conserv Genet 5:585–602
Pearson WR, Lipman DJ (1988) Improved tools for biological sequence comparison. Proc Natl Acad Sci USA 85:2444–2448
Penny D, Lockhart PJ, Steel MA, Hendy MD (1994) The role of models in reconstructing evolutionary trees. In: Scotland RW, Siebert DJ, Williams DM (eds) Models in phylogenetic reconstruction. Clarendon, Oxford, pp 211–230
Pérez-Losada M, Viscidi RP, Demma JC, Zenilman J, Crandall KA (2005) Population genetics of *Neisseria gonorrhoeae* in a high prevalence community using a hyper-variable outer membrane porB and thirteen slowly-evolving housekeeping genes. Mol Biol Evol 22:1887–1902
Pérez-Losada M, Browne EB, Madsen A, Wirth T, Viscidi RP, Crandall KA (2006) Population genetics of microbial pathogens estimated from multilocus sequence typing (MLST) data. Infect Genet Evol 6:97–112
Pond SL, Frost SD (2005) A genetic algorithm approach to detecting lineage-specific variation in selection pressure. Mol Biol Evol 22:478–485
Posada D, Buckley TR (2004) Model selection and model averaging in phylogenetics: advantages of Akaike information criterion and Bayesian approaches over likelihood ratio tests. Syst Biol 53:793–808

Posada D, Crandall KA (1998) MODELTEST: testing the model of DNA substitution. Bioinformatics 14:817–818
Posada D, Crandall KA (2001) Evaluation of methods for detecting recombination from DNA sequences: computer simulations. Proc Natl Acad Sci USA 98:13757–13762
Posada D, Crandall KA (2002) The effect of recombination on the accuracy of phylogeny estimation. J Mol Evol 54:396–402
Posada D, Crandall KA, Templeton AR (2000) GeoDis: a program for the cladistic nested analysis of the geographical distribution of genetic haplotypes. Mol Ecol 9:487–488
Posada D, Crandall KA, Holmes EC (2002) Recombination in evolutionary genomics. Annu Rev Genet 36:75–97
Pupko T, Huchon D, Cao Y, Okada N, Hasegawa M (2002) Combining multiple data sets in a likelihood analysis: Which models are the best? Mol Biol Evol 19:2294–2307
Pybus OG, Rambaut A (2002) GENIE: estimating demographic history from molecular phylogenies. Bioinformatics 18:1404–1405
Raftery AE (1996) Hypothesis testing and model selection. In: Gilks WR, Richardson S, Spiegelhalter DJ (eds) Markov chain Monte Carlo in practice. Chapman & Hall, London, pp 163–188
Rambaut A, Drummond AJ (2003) Tracer version 1.2.1. http:\evolve.zoo.ox.ac.uk.
Redelings BD, Suchard MA (2005) Joint Bayesian estimation of alignment and phylogeny. Syst Biol 54:401–418
Rivera MC, Lake JA (2004) The ring of life provides evidence for a genome fusion origin of eukaryotes. Nature 431:152–155
Ronquist F, Huelsenbeck JP (2003) MrBayes 3: Bayesian phylogenetic inference under mixed models. Bioinformatics 19:1572–1574
Rosenberg MS (2005) Evolutionary distance estimation and fidelity of pair wise sequence alignment. BMC Bioinformatics 6:102
Rzhetsky A, Nei M (1993) Theoretical foundation of the minimum-evolution method of phylogenetic inference. Mol Biol Evol 10:1073–1095
Sainudiin R, Wong WS, Yogeeswaran K, Nasrallah JB, Yang Z, Nielsen R (2005) Detecting site-specific physicochemical selective pressures: applications to the class I HLA of the human major histocompatibility complex and the SRK of the plant sporophytic self-incompatibility system. J Mol Evol 60:315–326
Saitou N, Nei M (1987) The neighbor-joining method: a new method for reconstructing phylogenetic trees. Mol Biol Evol 4:406–425
Sanderson MJ (1995) Objections to bootstrapping phylogenies: a critique. Syst Biol 44:299–320
Sanderson MJ, Wojciechowski MF (2000) Improved bootstrap confidence limits in large-scale phylogenies, with an example from Neo-Astragalus (Leguminosae). Syst Biol 49:671–685
Schierup MH, Hein J (2000) Consequences of recombination on traditional phylogenetic analysis. Genetics 156:879–891
Schmidt HA, Strimmer K, Vingron M, von Haeseler A (2002) TREE-PUZZLE: maximum-likelihood phylogenetic analysis using quartets and parallel computing. Bioinformatics 18:502–504
Schneider S, Roessli D, Excoffier L (2000) ARLEQUIN version 2.000: a software for population genetics data analysis. University of Geneva
Schwarz G (1978) Estimating the dimensions of a model. Ann Stat 6:461–464
Shimodaira H (2002) An approximately unbiased test of phylogenetic tree selection. Syst Biol 51:492–508
Shimodaira H, Hasegawa M (1999) Multiple comparisons of log-likelihoods with applications to phylogenetic inference. Mol Biol Evol 16:1114–1116
Shimodaira H, Hasegawa M (2001) CONSEL: for assessing the confidence of phylogenetic tree selection. Bioinformatics 17:1246–1247
Shriner D, Nickle DC, Jensen MA, Mullins JI (2003) Potential impact of recombination on site-wise approaches for detecting positive natural selection. Genet Res 81:115–121

Simes RJ (1986) An improved Bonferroni procedure for multiple tests of significance. Biometrika 73:751–754
Simon D, Larget B (2000) Bayesian analysis in molecular biology and evolution (BAMBE) version 2.03 beta. Duquesne University
Sinclair EA, Pérez-Losada M, Crandall K.A (2005) Molecular phylogenetics for conservation biology. In: Purvis A, Brooks T, Gittleman J (ed) Phylogeny and conservation. Cambridge University Press, Cambridge, pp 19–56
Sokal RR, Sneath PHA (1963) Numerical taxonomy. Freeman, San Francisco
Stamatakis AT, Ludwig H, Meier H (2005) RAxML-III: a fast program for maximum likelihood-based inference of large phylogenetic trees. Bioinformatics 21:456–463
Stephens M (2001) Inference under the coalescent. In: Balding DJ, Cannings C, Bishop M (eds) Handbook of statistical genetics. Wiley, Chichester, pp 213–238
Strimmer K, Pybus OG (2001) Exploring the demographic history of DNA sequences using the generalized skyline plot. Mol Biol Evol 18:2298–2305
Strimmer K, Rambaut A (2002) Inferring confidence sets of possibly misspecified gene trees. Proc Biol Sci 269:137–142
Suchard MA, Weiss RE, Sinsheimer JS (2001) Bayesian selection of continuous-time Markov chain evolutionary models. Mol Biol Evol 18:1001–1013
Sullivan J, Swofford DL (1997) Are guinea pigs rodents? The importance of adequate models in molecular phylogenies. J Mamm Evol 4:77–86
Sullivan J, Joyce P (2005) Model selection in phylogenetics. Annu Rev Ecol Evol Syst 36:445–466
Sullivan J, Markert JA, Kilpatrick CW (1997) Phylogeography and molecular systematics of the *Peromyscus aztecus* species group (Rodentia: Muridae) inferred using parsimony and likelihood. Syst Biol 46:426–440
Suzuki Y, Gojobori T (1999) A method for detecting positive selection at single amino acid sites. Mol Biol Evol 16:1315–1328
Suzuki Y, Gojobori T, Nei M (2001) ADAPTSITE: detecting natural selection at single amino acid sites. Bioinformatics 17:660–661
Suzuki Y, Glazko GV, Nei M (2002) Overcredibililty of molecular phylogenies obtained by Bayesian phylogenetics. Proc Natl Acad Sci USA 99:16138–16143
Swofford DL (2002) PAUP*. Phylogenetic analysis using parsimony (*and other methods). Sinauer, Sunderland
Swofford DL, Olsen GJ, Waddell PJ, Hillis DM (1996) Phylogenetic inference. In: Hillis DM, Moritz C, Mable BK (eds) Molecular systematics. Sinauer, Sunderland, pp 407–514
Tamura K, Nei M, Kumar S (2004) Prospects for inferring very large phylogenies by using the neighbor-joining method. Proc Natl Acad Sci USA 101:11030–11035
Taylor SD, de la Cruz KD, Porter ML, Whiting MF (2005) Characterization of the long-wavelength opsin from Mecoptera and Siphonaptera: does a flea see? Mol Biol Evol 22:1165–1174
Templeton AR (1998) Nested clade analyses of phylogeographic data: testing hypotheses about gene flow and population history. Mol Ecol 7:381–397
Templeton AR (2004) Statistical phylogeography: methods of evaluating and minimizing inference errors. Mol Ecol 13:789–810
Templeton AR, Crandall KA, Sing CF (1992) A cladistic analysis of phenotypic associations with haplotypes inferred from restriction endonuclease mapping and DNA sequence data. III. Cladogram estimation. Genetics 132:619–633
Templeton AR, Sing CF (1993) A cladistic analysis of phenotypic associations with haplotypes inferred from restriction endonuclease mapping. IV. Nested analyses with cladogram uncertainty and recombination. Genetics 134:659–669
Thompson JD, Higgins DG, Gibson TJ (1994) CLUSTAL W: improving the sensitivity of progressive multiple sequence alignment through sequence weighting, position-specific gap penalties and weight matrix choice. Nucleic Acids Res 22:4673–4680

Thompson JD, Gibson TJ, Plewniak F, Jeanmougin F, Higgins DG (1997) The CLUSTAL_X windows interface: flexible strategies for multiple sequence alignment aided by quality analysis tools. Nucleic Acids Res 25:4876–4882
Thompson JD, Plewniak F, Poch O (1999a) BAliBASE: a benchmark alignment database for the evaluation of multiple alignment programs. Bioinformatics 15:87–88
Thompson JD, Plewniak F, Poch O (1999b) A comprehensive comparison of multiple sequence alignment programs. Nucleic Acids Res 27:2682–2690
Tibayrenc M (2005) Bridging the gap between molecular epidemiologists and evolutionists. Trends Microbiol 13:575–580
Tierney L (1994) Markov-chains for exploring posterior distributions. Ann Stat 22:1701–1728
Urwin R, Maiden MC (2003) Multi-locus sequence typing: a tool for global epidemiology. Trends Microbiol 11:479–487
Wain-Hobson S, Renoux-Elbe C, Vartanian JP, Meyerhans A (2003) Network analysis of human and simian immunodeficiency virus sequence sets reveals massive recombination resulting in shorter pathways. J Gen Virol 84:885–895
Wall JD (2000) A comparison of estimators of the population recombination rate. Mol Biol Evol 17:156–163
Watterson GA (1975) On the number of segregating sites in genetical models without recombination. Theor Popul Biol 7:256–276
Wernersson R, Pedersen AG (2003) RevTrans: multiple alignment of coding DNA from aligned amino acid sequences. Nucleic Acids Res 31:3537–3539
Wheeler WC (1996) Optimization alignment: the end of multiple alignment in phylogenetics? Cladistics 12:1–9
Wheeler WC, Gladstein D, De Laet J (2003) POY: phylogeny reconstruction via optimization of DNA and other data. American Museum of Natural History, New York. http://research.amnh.org/scicomp/projects/poy.php
Whelan S, Goldman N (2001) A general empirical model of protein evolution derived from multiple protein families using a maximum-likelihood approach. Mol Biol Evol 18:691–699
Williamson S, Orive ME (2002) The genealogy of a sequence subject to purifying selection at multiple sites. Mol Biol Evol 19:1376–1384
Wilson DJ, McVean G (2006) Estimating diversifying selection and functional constraint in the presence of recombination. Genetics 172:1411–1425
Wong WS, Yang Z, Goldman N, Nielsen R (2004) Accuracy and power of statistical methods for detecting adaptive evolution in protein coding sequences and for identifying positively selected sites. Genetics 168:1041–1051
Wooding S, Rogers A (2002) The matrix coalescent and an application to human single-nucleotide polymorphisms. Genetics 161:1641–1650
Woolley S, Johnson J, Smith MJ, Crandall KA, McClellan DA (2003) TreeSAAP: selection on amino acid properties using phylogenetic trees. Bioinformatics 19:671–672
Xia X, Li WH (1998) What amino acid properties affect protein evolution? J Mol Evol 47:557–564
Yang Z (1997) PAML: a program package for phylogenetic analysis by maximum likelihood. Comput Appl Biosci 13:555–556
Yang Z, Bielawski JP (2000) Statistical methods for detecting molecular adaptation. Trends Ecol Evol 15:496–503
Yang Z, Nielsen R (2000) Estimating synonymous and nonsynonymous substitution rates under realistic evolutionary models. Mol Biol Evol 17:32–43
Yang Z, Nielsen R (2002) Codon-substitution models for detecting molecular adaptation at individual sites along specific lineages. Mol Biol Evol 19:908–917
Yang Z, Rannala B (2005) Branch-length prior influences Bayesian posterior probabilities of phylogeny. Syst Biol 54:455–470
Yang Z, Swanson WJ (2002) Codon-substitution models to detect adaptive evolution that account for heterogeneous selective pressures among site classes. Mol Biol Evol 19:49–57

Yang Z, Nielsen R, Goldman N, Pedersen AM (2000) Codon-substitution models for heterogeneous selection pressure at amino acid sites. Genetics 155:431–449

Yang Z, Wong WS, Nielsen R (2005) Bayes empirical Bayes inference of amino acid sites under positive selection. Mol Biol Evol 22:1107–1118.

Zhang J, Nei M (1997) Accuracies of ancestral amino acid sequences inferred by the parsimony, likelihood, and distance methods. J Mol Evol 44:S139–S146

Zwickl DJ, Holder MT (2004) Model parameterization, prior distributions, and the general time-reversible model in Bayesian phylogenetics. Syst Biol 53:877–888

Chapter 11
Virus Evolution and Taxonomy

Anne-Lise Haenni

Abstract With the demonstration of the existence of viruses over 100 years ago, and the plethora of different viruses subsequently described in all types of cells, it became urgent to classify these microorganisms. The first efforts in this direction were initiated in 1966, and in 1973 the International Committee on Taxonomy of Viruses was created. The task of this committee is to classify viruses hierarchically, using the four main taxa, order, family, genus and species. However, classification is rendered difficult because major elements drive virus evolution, such as mutation, recombination and reassortment. Plant viruses are very well suited to establish a means of virus classification and to study virus evolution, given the ease with which large numbers of plants can be infected and high yields of virus rapidly obtained. In addition, using plant viruses obviates the restrictions generally linked to animal experimentation. Examples are provided among plant viruses illustrating the extent to which some of the elements driving evolution have participated in structuring the viruses known today, and hence the complexity there is in constructing phylogenetic trees of viruses. Consequently, grouping viruses in a durable taxonomic system is particularly hazardous and classification must proceed with caution.

Anne-Lise Haenni
Institut Jacques Monod, 2 Place Jussieu Tour 43, 75251 Paris Cedex 05, France
haenni@ijm.jussieu.fr

M.J. Roossinck (ed.), *Plant Virus Evolution.*

11.1 Brief Overview of Virus Taxonomy

The classification and naming of organisms were attempted several centuries ago in Europe, but at the time did not provide a simple workable system. However, as a result of more systematic studies and of the arrival of numerous foreign organisms brought to Europe from other continents, it became urgent to adopt a simplified naming system. This was first established and developed by the Swedish physician and botanist Linnaeus, in *Species Plantarum*, in 1753, and was applied first to plants and thereafter to animals. Linnaeus introduced the binomial system comprising a Latin name for the genus and another for the species. He was the first to use the binomial system consistently, and this system still constitutes the basis of organism classification today.

With the discovery of viruses towards the end of the nineteenth century and the ensuing demonstration that viruses exist in virtually all types of living organisms, it also appeared important to develop a system of classification of viruses. As a result, viruses – including plant viruses – were initially classified into groups on the basis of similarities such as particle morphology. Plant viruses were at times also classified on the basis of their cross-protection capacity, i.e., the capacity of a mild virus to infect and subsequently protect plants from infection by a related severe virus (discussed in Roossinck 2005).

The first international endeavour towards virus classification was the creation of the International Committee on Nomenclature of Viruses in 1966. In 1973, this committee became known as the International Committee on Taxonomy of Viruses (ICTV; Mayo and Pringle 1998). Some time elapsed, however, before it was agreed that species, genera and higher taxa were appropriate for the classification of viruses, and only in 1989 did the ICTV adopt a definition of a species that incorporated the idea of relationship by descent. The ICTV's activities now include making regular reports of relevant information available to the international virology community. This is achieved mainly via its Web site (http://www.ictronline.org/) and by two types of publications: (1) in the Virology Division News section of *Archives of Virology* and (2) regular reports on the state of virus taxonomy; the eighth such report (Fauquet et al. 2005) appeared in 2005.

11.2 Elements Dictating Taxonomy

Virus taxonomy is a means of classifying viruses that can be defined as (1) the arranging (classification) of viruses into related clusters, (2) the identification of the extent of relatedness within and between these clusters, and (3) the giving of names (nomenclature) to these clusters, also known as taxa (Mayo and Pringle 1998). The ideal situation would be one in which a permanent taxonomic system could be worked out. However, with the appearance of new viruses, as well as with the development of more refined methods of analysis that have revealed new gene arrangements and new relationships among viruses belonging to different groups, it is clear that such an ideal situation does not exist. Hence, taxonomy must continuously adapt to new information as it is acquired. Consequently, taxonomy must advance with caution, refraining from taking steps that could result from having weighed possible alternatives poorly. Such a conservative attitude has inevitably led to a certain level of controversy among virologists (Van Regenmortel 2007).

Taxonomy implies that organisms are classified hierarchically such that their mutual relationship is expressed in a simple manner that complies with internationally agreed codes of practice (Haenni and Mayo 2006). For virus taxonomy to be credible, it should comply with the principles of stability, utility, acceptability and flexibility (Mayo and Pringle 1998).

The hierarchical classification of viruses as recognized currently by the ICTV, uses four main taxa, order, family, genus and species; intermediate taxa such as subfamilies are rarely used. The ICTV does not classify viruses below the level of species. In addition, some genera are not classified in families because the information available is not sufficient to determine how these viruses might best be grouped to form higher taxa. Among virus species, there are many that cannot be inserted appropriately into an existing genus, and they therefore remain as tentative species.

One of the driving forces in the construction of virus taxonomy has been the study of phylogeny. In recent years, the study of phylogeny has been greatly facilitated by the availability of rapid genome sequencing methods, of methods to compare protein structures and of programs for the deduction of phylogenies. Because virus genomes are small, they are readily amenable to such studies. Phylogenetic studies have thus prompted considerations of the origin of viruses and of virus evolution.

11.3 Taxonomy and Evolution

The main elements that drive virus evolution are point mutations, recombination and reassortment. These are basically similar to those that drive all evolution, since mechanisms of recombination also prevail during cell meiosis, and reassortment also operates during chromosome redistribution in sexual reproduction. Such forces lead to diversity among virus genomes and these genomes are then subject to natural selection.

11.3.1 Point Mutations

Such mutations generally occur during virus genome replication. They are particularly frequent among RNA viruses, owing to the low fidelity of their RNA-dependent RNA polymerases (RdRps) that are not endowed with proofreading capacity.

Editing is another mechanism that can introduce errors in RNA-containing viral genomes. It would seem logical in these circumstances, to assume that the frequency of mutations would be similar for all viruses containing an RNA genome, and would occur anywhere along the viral genome. However, this is not what is observed, as demonstrated for *Human immunodeficiency virus 1* (reviewed in Rambaut et al. 2004). The reasons for this might lie in the fact that (1) even if mutations were to occur randomly, some of the resulting progeny viruses may not be viable and would be eliminated from the genetic pools that constitute viruses and (2) specific structures of the virus RNA, or the binding of proteins to specific sites on the RNA, might cause the RdRp to pause, and this could in turn result in major variations in the genome at such sites (see Chap. 6 for more details on mutations).

11.3.2 Recombination

This necessarily involves two genomes coexisting in the same cell. It refers to the joining of parts of two separate RNA molecules to form a molecule with a new sequence. Several experiments have demonstrated that recombination – which occurs during virus genome replication in particular among RNA viruses – has led to the acquisition of genes or blocks of genes from other viral genomes, a phenomenon known as "modular evolution" in which functional modules from different virus sources participate in creating new viruses (reviewed in Worobey and Holmes 1999). In certain viruses such as those of the family *Luteoviridae* (see later), putative recombination events appear to occur at specific sites in the virus genome, known as "hot spots" that coincide with initiation sites for subgenomic RNA transcription (Miller et al. 1995, 1997). Recombination generally occurs by a template-switch mechanism. This is probably the area of virus evolution that has benefited the most from the advent of new tools in molecular biology such as the development of PCR. In particular, sequence analyses and phylogenetic techniques have led to the detection of multiple recombination events in nature (reviewed in Worobey and Holmes 1999; Aaziz and Tepfer 1999; see Chap. 8). Three major types of recombination can be distinguished; these are homologous recombination, aberrant recombination and nonhomologous recombination. The precise mechanisms involved in the various types of recombination have been recently examined using various plant viruses such as *Turnip crinkle virus* (genus *Carmovirus*, family *Tombusviridae*; Nagy et al. 1998) and *Brome mosaic virus* (BMV; genus *Bromovirus*, family *Bromoviridae*; Alejska et al. 2001).

11.3.2.1 Homologous Recombination

The RNA molecules are related, and crossing-over by the RdRp from one RNA to the other occurs at homologous or comparable regions in the two molecules. This can allow two viral genomes with different deleterious mutations to regenerate a functional genome or a genome better adapted to its environment. Such events can also occur between a cellular transgene expressing part of a virus genome and the corresponding virus genome harboring a gene defect, as demonstrated in the case of *Cauliflower mosaic virus*, and also between the messenger RNAs of certain transgenes and the RNA genomes of certain viruses (reviewed in Miller et al. 1997).

11.3.2.2 Aberrant Recombination

In this case, crossing-over between two related RNA molecules occurs at unrelated sites, although generally close to related sites. This leads to duplications or deletions in the ensuing progeny, and can also result in the introduction of nucleotides of unknown origins. This is unique to RNA viruses and is common when defective RNAs are involved.

11.3.2.3 Nonhomologous Recombination

This occurs when crossing-over takes place between genome sequences that share no homology. It is responsible for gene rearrangements, insertions and deletions. It is also responsible for the introduction into virus genomes of genes of host cell origin; such situations have been described in some preparations of a luteovirus isolate whose 5′-terminal sequence is derived from a chloroplast exon (Mayo and Jolly 1991) and among closteroviruses that have acquired host cell protein-coding regions (Dolja et al. 1994).

11.3.3 Reassortment

This involves the exchange of genome segments between two viruses with a segmented genome that coexist in the same cell. It has played a crucial role in the evolution of plant viruses containing segmented genomes, such as that of *Cucumber mosaic virus* (genus *Cucumovirus*, family *Bromoviridae*) and also in the genus *Tobravirus* (reviewed in Roossinck 2002), and in the family *Comoviridae* (De Jager 1976).

The mosaicism that results in particular from recombination and reassortment leads to tremendous difficulties in establishing a possible and durable taxonomic classification system based on sequence comparisons; it demonstrates the necessity of flexibility when dealing with taxonomy.

In addition it should be stressed that the RNA genomes of viruses are composed of a cloud (or swarm) of closely related molecules known as "quasispecies" (Eigen 1996; Biebricher and Eigen 2006; Domingo et al. 2006) that allows the virus to adapt to environmental changes and survive in front of host species that are evolving.

Should the quasispecies spectrum decrease such as in bottleneck conditions (Li and Roossinck 2004), the fitness of the virus population may also decrease. Virus–host as well as virus–vector interactions (in particular insect vectors when the virus is harbored in a persistent manner) are important in promoting virus diversity and evolution, and for the survival of the virus.

11.4 The Problems of Plant Virus Taxonomy

Plant viruses are particularly well suited to the study of virus evolution and the experimental verification of hypotheses concerning the mechanisms responsible for virus evolution: a large number of plants can be easily infected, the viruses introduced multiply rapidly to high levels, and they can easily be analyzed by current methods. It is easy to produce a large number of transgenic or infected plants, and such experiments are not subjected to the same severe constraints that dictate experimentation on animals. In addition, plants can easily be resampled. Many model plants can be reproduced rapidly, within a matter of weeks. Hence, the evolution of plant viruses has become an important area of research.

Among RNA-containing viruses, phylogenetic trees have frequently been based on the sequences of the RdRps that can be compared across distinct genera and families, because of the similarities that exist among them. It is through such analyses that links have been detected between certain plant and animal viruses. However, complications arise if phylogenetic trees are constructed on the basis of a single gene, and single phylogenetic trees are insufficient to describe evolutionary relationships among viruses (Worobey and Holmes 1999). In addition, alignments, models of evolution and specific methods must be carefully evaluated to ensure an accurate phylogenetic estimation (see Chap. 10). To reach credible conclusions, the entire genomes rather than short sequence regions should be compared so as to take into account all the phylogenetic relationships that are possible (Zanotto et al. 1996).

Several virus groups have been examined in considerable detail from the point of view of their evolution, and the results highlight the importance of recombination in forging these viruses as we know them today. A few such examples will be considered here. They reveal that genomic mosaic structures occur in plant viruses and are the result of frequent recombination events.

11.4.1 Members of the Luteoviridae *and Their Affiliation to Distinct Genera*

The family *Luteoviridae* currently includes three genera, *Luteovirus*, *Polerovirus* and *Enamovirus*. These are distinct from the genera *Sobemovirus* and *Umbravirus*, neither of which is included in a family (D'Arcy and Mayo 1997; Fauquet et al. 2005).

Members of the family *Luteoviridae* provide good examples of the involvement of genome parts in recombination and illustrate the problems that arise in relating taxonomy to evolution.

Poinsettia latent virus (PnLV) has been classified as a tentative species in the genus *Alphacryptovirus* (family *Partitiviridae*; Fauquet et al. 2005). Recent characterization of the virus has revealed that it is a chimera (Aus dem Siepen et al. 2005). The 5′-terminal three quarters of its monopartite, single-stranded RNA genome appears to be derived from the genome of a polerovirus, whereas the 3′-terminal quarter is related to that in the genomes of members of the genus *Sobemovirus* (Aus dem Siepen 2005; Fig. 11.1). Hence, open reading frame (ORF) 1 and ORFs 2 and 3 that encode the replication functions of PnLV probably originate from a polerovirus, whereas ORF 4 that codes for the coat protein probably originates from a sobemovirus; thus PnLV, a recombinant, would have crossed family borders. In addition, the last ten nucleotides at the 5′ end as well as the last four nucleotides at the 3′ end of PnLV are identical to the sequences found in the same positions in the genomes of poleroviruses, suggesting that further links with poleroviruses possibly exist.

Sugarcane yellow leaf virus (ScYLV) is an emerging virus that causes yellow leaf syndrome with severe economic consequences for the sugarcane industry. On the basis of sequence analyses of isolates of this virus (Maia et al. 2000; Moonan et al. 2000; Smith et al. 2000; reviewed in Roossinck 2005), it has been concluded that ScYLV probably arose by recombination between members of the three genera that compose the family *Luteoviridae*; indeed ORFs 1 and 2 of the ScYLV-Florida (ScYLV-F) isolate are most similar to the corresponding ORFs of poleroviruses, whereas ORFs 3 and 4 present the highest similarity to ORFs 3 and 4 of luteovirus sequences, and ORF 5 is most closely related to the read-through protein gene of the *Enamovirus*, *Pea enation mosaic virus-1* (Smith et al. 2000). Thus, ScYLV-F seems to harbor regions homologous to all three genera of the *Luteoviridae*: the 5′ part of its genome might have originated from a polerovirus, whereas the 3′ region would be related to the genome of a luteovirus and of an enamovirus. By extending such comparisons further, Moonan et al. (2000) showed that the start site for the synthesis of subgenomic RNA (sgRNA) 1 (and of sgRNA 2 in *Barley yellow dwarf virus*; genus *Luteovirus*) corresponds to the site of recombination

A related situation has been reported for *Soybean dwarf virus* (SbDV; classified as a member of the *Luteovirus*) The 5′ half of its genome bears apparent sequence homology with the genome of members of the *Luteovirus*, and its 3′ half with the genome of members of the *Polerovirus*. SbDV may therefore have arisen by intergenic recombination between polerovirus and luteovirus ancestors (Rathjen et al. 1994).

Thus, interspecies recombination events have occurred frequently during evolution in this family of viruses. This raises the question of which is the parental lineage and which is the recombinant. If one recombinational event can be detected, no clear answer to this question can be provided. However, if more than one event is detected, such as the two events that probably occurred in ScYLV-F, it is likely that the virus is a recombinant with sequences from a polerovirus, a luteovirus and an

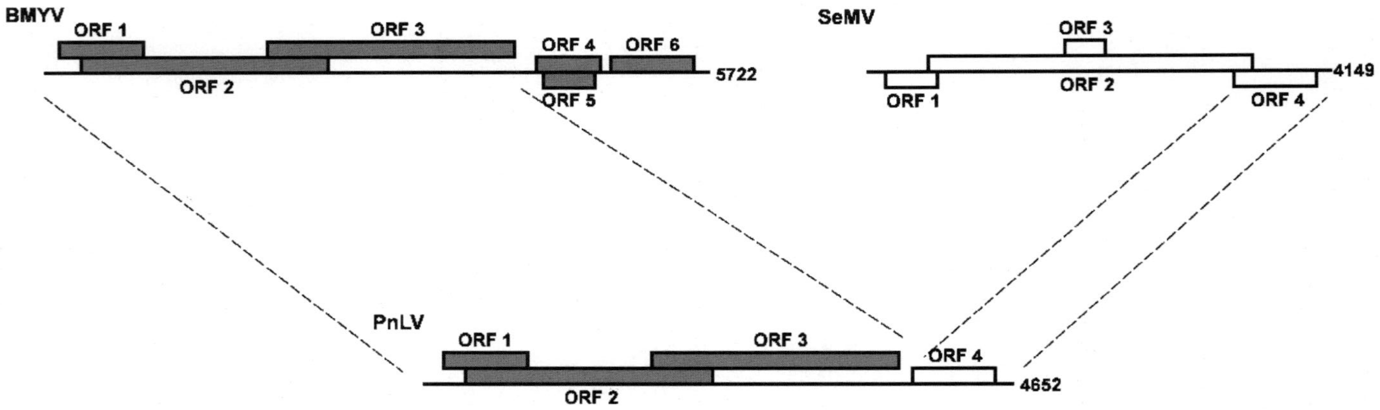

Fig. 11.1 Comparison of the genome map of *Poinsettia latent virus* (*PnLV*) with that of *Beet mild yellowing virus* (*BMYV*; *Polerovirus*) and that of *Sesbenia mosaic virus* (*SeMV*; *Sobemovirus*). The *horizontal line* represents the genome; *the number to the right of the line* indicates the length of the genome in nucleotides. *Boxes* correspond to open reading frames (*ORF*) and are numbered from the 5′-terminal end (left) to the 3′-terminal end (right). Similar shadings of ORFs indicated strong relationships. Not to scale. (Adapted from Aus dem Siepen et al. 2006)

enamovirus. These examples reveal the difficulties encountered in attempting to relate taxonomy to apparent evolutionary pathways: the resulting taxonomy depends on the virus proteins considered for classification.

11.4.2 Transfer-RNA-Like Structures in Viral RNAs

It has been known for several decades that the 3′ end of the noncoding region of the RNA genome of viruses of certain genera or families can be specifically aminoacylated by an aminoacyl transfer RNA (tRNA) synthetase. The amino acids that can be bound to such viral RNAs are valine, histidine and tyrosine, the amino acid depending on the viral genome. Originally, the occurrence of a tRNA-like structure was demonstrated by the aminoacylation capacity of the viral RNA: valylation (Yot et al. 1970; Pinck et al. 1970) of *Turnip yellow mosaic virus* (TYMV; genus *Tymovirus*, family *Tymoviridae*) RNA, histidylation (Oberg and Philipson 1972) of *Tobacco mosaic virus* (TMV; genus *Tobamovirus*) RNA and tyrosylation (Hall et al. 1972) of BMV RNA. This led to the study of the sequences composing the tRNA-like regions, and to their presumed structures. With the advent of sequencing methods and software that allow more direct analyses of the possible folding of the 3′-terminal regions of viral RNAs to be performed, together with the features of already established tRNA-like structures and of tRNAs, it became possible to predict the presence of such tRNA-like structures in newly sequenced viral RNAs, and subsequently to validate the folding pattern by possible aminoacylation reactions. This approach also initiated the study of pseudoknots, which are now known to be primordial in most RNA folding patterns.

It is now established that a viral RNA with a given overall tRNA-like structure specifically binds a given amino acid (valine, histidine or tyrosine). Examination of the features of tRNA-like structures has led to the classification of the viruses possessing such structures in group A or B. Viruses of group A carry a simple tRNA-like structure, whereas those of group B are far more elaborate.

Basically, and disregarding a few exceptions (some of which are discussed below), the situation regarding the aminoacylation capacity of viral RNAs can be summarized as in Table 11.1. In addition, the group A tRNA-like structure of TMV and of other members of the genus *Tobamovirus* that accept histidine is preceded by a region forming an upstream pseudoknot domain (UPD). In some cases, a hairpin (HP) structure is also located between the UPD and the tRNA-like structure.

Recent results (Dreher 1999; Fechter et al. 2001; Koenig et al. 2005; Rudinger-Thirion et al. 2006) obtained by examining the genomes of viruses possessing a tRNA-like structure and scrutinizing their encoded proteins, from sequence, structural and functional points of view, have led to the suggestion that the genomes have undergone considerable recombination events. To this must be added other phenomena such as point mutations, possible bottlenecks and selection of the fittest variants. Thus, the situation is not as simple as might appear at first sight.

Table 11.1 Summary of the aminoacylation capacity of plant virus RNAs

Group	Genus	Virus	Aminoacylation
A	*Tymovirus*	*Turnip yellow mosaic virus*	Valine
	Furovirus	*Soil-borne wheat mosaic virus*	Valine
	Pomovirus	*Beet soil borne virus*	Valine
	Pecluvirus	*Peanut clump virus 1*	Valine
	Tobamovirus	*Tobacco mosaic virus*	Histidine
B	*Bromovirus*	*Brome mosaic virus*	Tyrosine
	Hordeivirus	*Barley stripe mosaic virus*	Tyrosine

A virus is provided as an example for each genus.

One striking example concerns the aminoacylation of a *Tymovirus*, *Nemesia ring necrosis virus* (NeRNV), and that of a *Tobamovirus*, *Sunhemp mosaic virus* (SHMV, previously CcTMV). Whereas the RNAs of most members of the *Tymoviridae* can be valylated, that of NeRNV is histidylated (Koenig et al. 2005); similarly, whereas the RNAs of members of the *Tobamovirus* are generally histidylated, the RNA of SHMV is valylated (Meshi et al. 1981). This is schematized in Fig. 11.2. This figure also shows that the 3′ untranslated region of these virus RNAs can contain (1) a tRNA-like structure, (2) an UPD followed by a tRNA-like structure or (3) an UPD followed by a HP and by a tRNA-like structure. Several recombination events followed by selection of variants have probably led to this diversity of 3′ ends.

11.5 Concluding Remarks

The examples presented illustrate cases wherein recombination can involve large chunks of viral genomes. Blocks such as genes may then have a phylogeny that can be presented in a conventional tree. Using the entire genome for a phylogenetic analysis may be far more difficult and misleading, in particular when different sections of the genome clearly arrived from different routes. These complex interrelationships represent a challenge to virologists: to develop methods for the presentation of these relationships. Owing to the recombinative nature of viruses, no single phylogenetic tree can account for the evolutionary relatedness that exists between viruses (Worobey and Holmes 1999). New methods that describe relationships as networks rather than trees (see Chap. 10) may hold the key to understanding these relationships. Trying to fit these relationships into a two-dimensional taxonomic hierarchy is clearly illusory.

These examples also illustrate the complexity of establishing a durable and comprehensive taxonomic system for viruses, and the risks that arise in shifting viruses from one group to another without sufficient background information. Viruses as classified today for the most part have undergone considerable shuffling of their

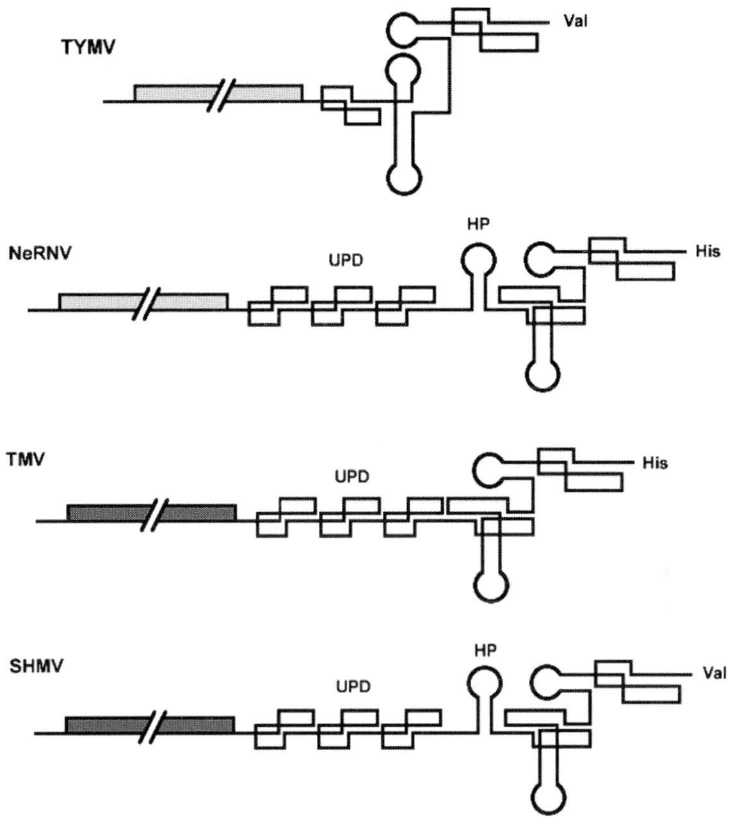

Fig. 11.2 The 3′ termini of the aminoacylatable plant viral RNAs discussed in the text, and of the amino acid that can be bound to the 3′ end of the RNA genome. *Boxes* correspond to ORFs; *light gray boxes* correspond to the ORFs of viruses of the genus *Tymovirus* and *dark gray boxes* correspond to ORFs of viruses of the genus *Tobamovirus*. The transfer RNA (tRNA) like structure occupies the 3′ end of the viral genome. *UPD* corresponds to the series of two or more pseudoknot structures located upstream of the tRNA-like structure. *HP* corresponds to a hairpin observed in the genome of certain viruses, between the UPD and the tRNA-like structure. Not to scale. *TYMV, Turnip yellow mosaic virus*; *NeRNV, Nemesia ring necrosis virus*; *TMV, Tobacco mosaic virus*; *SMHV, Sunhemp mosaic virus*. (Adapted from Koenig et al. 2005 and Rudinger-Thirion et al. 2006)

genome parts over the years. The availability of refined techniques of analysis of virus genomes should now allow important advances to be made in our understanding of virus evolution.

In summary, taxonomists should proceed slowly towards classification and once classification has been achieved, they should defend existing schemes in the interests of nomenclatural stability, but should be ready to adapt their systems to accommodate new findings and apparently heretical views derived from them.

Acknowledgements The author is very grateful to Michael J. Adams and Mike A. Mayo for helpful suggestions and constructive comments, and wishes to thank Carlos Giraldo for his help with the figures. The author gratefully acknowledges the support of the Centre National de la Recherche Scientifique, Paris, France, and of Colciencias, University of Antioquia, Medellin, Colombia.

References

Aaziz R, Tepfer M (1999) Recombination in RNA viruses and in virus-resistant transgenic plants. J Gen Virol 80:1339–1346

Alejska M, Kurzynska-Kokorniak A, Broad M, Kierzek R, Figlerowicz M (2001) How RNA viruses exchange their genetic material. Acta Biochim Pol 48:391–407

Aus dem Siepen M, Pohl JO, Koo B-J, Wege C, Jeske H (2005) Poinsettia latent virus is not a cryptic virus, but a natural polerovirus-sobemovirus hybrid. Virology 336:240–250

Biebricher CK, Eigen M (2006) What is a quasispecies? Curr Top Microbiol Immunol 299:1–31

D'Arcy CJ, Mayo M (1997) Proposals for changes in luteovirus taxonomy and nomenclature. Arch Virol 142:1285–1287

De Jager CP (1976) Genetic analysis of cowpea mosaic virus mutants by supplementation and reassortments tests. Virology 70:151–163

Dolja VV, Karasev AV, Koonin EV (1994) Molecular biology and evolution of closteroviruses: sophisticated build-up of large RNA genomes. Annu Rev Phytopathol 32:261–285

Domingo E, Martin V, Perales C, Grande-Perez A, Garcia-Arriaza J, Arias A (2006) Viruses as quasispecies: biological implications. Curr Top Microbiol Immunol 299:51–82

Dreher TW (1999) Functions of the 3′-untranslated regions of positive strand RNA viral genomes. Annu Rev Phytopathol 37:151–174

Eigen M (1996) On the nature of virus quasispecies. Trends Microbiol 4:216–218

Fauquet CM, Mayo MA, Malinoff J, Desselberger U, Ball LA (2005) Virus taxonomy. VIIIth report of the ICTV. Elsevier/Academic, London

Fechter P, Rudinger-Thirion J, Florentz C, Giege R (2001) Novel features in the tRNA-like world of plant viral RNAs. Cell Mol Life Sci 58:1547–1561

Haenni A-L, Mayo M (2006) Virus systematics: taxonomy for the tiny. Microbiol Today 156–159

Hall TC, Shih DS, Kaesberg P (1972) Enzyme-mediated binding of tyrosine to brome-mosaic-virus ribonucleic acid. Biochem J 129:969–976

Koenig R, Barends S, Gultyaev AP, Lesemann D-E, Vetten, HJ, Loss S, Pleij CWA (2005) Nemesia ring necrosis virus: a new tymovirus with a genomic RNA and a histidylatable tobamovirus-like 3′ end. J Gen Virol 86:1827–1833

Li H, Roossinck MJ (2004) Genetic bottlenecks reduce population variation in an experimental RNA virus population. J Virol 78:10582–10587

Maia IG, Goncalves MC, Arruda P, Veja J (2000) Molecular evidence that *Sugarcane yellow leaf virus* (ScYLV) is a member of the *Luteoviridae* family. Arch Virol 145:1009–1019

Mayo MA, Jolly CA (1991) The 5′-terminal sequence of potato leafroll virus RNA: evidence of recombination between virus and host RNA. J Gen Virol 72:2591–2595

Mayo MA, Pringle CR (1998) Virus taxonomy – 1997. J Gen Virol 79:649–657

Meshi T, Ohno T, Iba H, Okada Y (1981) Nucleotide sequence of a cloned cDNA copy of TMV (cowpea strain) RNA, including the assembly origin, the coat protein cistron, and the 3′ non-coding region. Mol Gen Genet 184:20–25

Miller WA, Dinesh-Kumar SP, Paul CP (1995) Luteovirus gene expression. Crit Rev Plant Sci 14:179–211

Miller WA, Koev G, Mohan BR (1997) Are there risks associated with transgenic resistance to luteoviruses? Plant Dis 81:700–710

Moonan F, Molina J, Mirkov TE (2000) Sugarcane yellow leaf virus: an emerging virus that has evolved by recombination between luteoviral and poleroviral ancestors. Virology 269:156–171

Nagy PD, Zhang C, Simon AE (1998) Dissecting RNA recombination in vitro: role of RNA sequences and the viral replicase. EMBO J 17:2392–2403

Oberg B, Philipson L (1972) Binding of histidine to tobacco mosaic virus RNA. Biochem Biophys Res Commun 48:927–932

Pinck M, Yot P, Chapeville F, Duranton HM (1970) Enzymatic binding of valine to the 3′ end of TYMV RNA. Nature 226:954–956

Rambaut A, Posada D, Crandall KA, Holmes EC (2004) The causes and consequences of HIV evolution. Nat Rev Genet 5:52–59

Rathjen JP, Karageorgos LE, Habili N, Waterhouse PM, Symons RH (1994) Soybean dwarf luteovirus contains the third variant genome type in the luteovirus group. Virology 198:671–679

Roossinck, MJ (2002) Evolutionary history of Cucumber mosaic virus deduced from phylogenetic analyses. J Virol 76:3382–3387

Roossinck MJ (2005) Symbiosis versus competition in plant virus evolution. Nat Rev Microbiol 3:917–924

Rudinger-Thirion J, Olsthoorn RCL, Giege R, Barends S (2006) Idiosyncratic behaviour of tRNA-like structures in translation of plant viral RNA genomes. J Mol Biol 355:873–878

Smith GR, Borg Z, Lockhart BE, Braithwaite KS, Gibbs MJ (2000) Sugarcane yellow leaf virus: a novel member of the Luteoviridae that probably arose by inter-species recombination. J Gen Virol 81:1865–1869

Van Regenmortel MHV (2007) Virus species and virus identification: past and current controversies. Infect Genet Evol 7:133–144

Worobey M, Holmes EC (1999) Evolutionary aspects of recombination in RNA viruses. J Gen Virol 80:2535–2543

Yot P, Pinck M, Haenni A-L, Duranton HM, Chapeville F (1970) Valine-specific tRNA-like structure in Turnip yellow mosaic virus RNA. Proc Natl Acad Sci USA 67:1345–1352

Zanotto PMA, Gibbs MJ, Gould EA, Holmes EC (1996) A reevaluation of the higher taxonomy of viruses based on RNA polymerases. J Virol 70:6083–6096

Index

Printing: Krips bv, Meppel, The Netherlands
Binding: Stürtz, Würzburg, Germany